NASA
Reference
Publication
1291

1993

Fractography Handbook of Spaceflight Metals

Rebecca J. Derro
Goddard Space Flight Center
Greenbelt, Maryland

National Aeronautics and
Space Administration

Office of Management

Scientific and Technical
Information Program

TABLE OF CONTENTS

PREFACE

This handbook was produced with the intention of providing failure analysts who work with spaceflight metals a reference of SEM fractographs of fracture surfaces produced under known conditions. The metals and the fracture conditions were chosen to simulate situations that are encountered in spaceflight applications. This includes tensile overload at both room temperature and liquid nitrogen temperature and fatigue at room temperature.

Overview

The handbook of fractography is used in failure analyses to help identify service failures. Comparisons of an unknown fracture surface to those of known failure mode and environmental conditions have proven helpful in the past. Although there are several such handbooks available, the need for a comprehensive book of scanning electron fractographs of the metals and temperature conditions important to Goddard Space Flight Center (GSFC) applications has become evident.

This book contains information on four common alloys used in spacecraft, including preparing samples for tensile overload and fatigue testing, preparing fracture specimens for examination in the scanning electron microscope (SEM), and documenting the fracture surfaces.

Materials Selection

Four alloys and a welded sample were selected for the handbook. These were 6061-T651 aluminum, 7075-T76 aluminum, copper beryllium-C17200-TF00, titanium-6 aluminum-4 vanadium, and welded 6061-T651 aluminum.

Each alloy has properties which make its use attractive to the space industry. Aluminum is used for structural applications because of its combination of strength and low density. Wrought aluminum alloys are identified with a four digit numerical designation. The first digit identifies the alloy group and the next three digits identify specifics such as aluminum purity and impurity limits. 2XXX, 6XXX, and 7XXX are designations for heat-treatable wrought aluminum alloys. The 6XXX and 7XXX series alloys are the most commonly used in the space industry. The major alloying elements for the 6XXX series are magnesium and silicon, and for the 7XXX series, zinc is the main alloying element. Alloys whose mechanical properties are modified by exposure to elevated temperatures are heat treatable. The heat treatments developed for these alloys are designated by the letter T with a numerical suffix.

The aluminum alloy, 7075, has copper, magnesium and zinc as its main alloying elements. T76 identifies the heat treatment applied to the 7075 aluminum alloy as solution heat-treated and stabilized. This alloy has high strength (tensile strength is 83 ksi at room temperature) and good corrosion resistance. 6061 is the designation for aluminum alloys containing magnesium and silicon as the main alloying elements. This alloy is used in lower strength applications (tensile strength is 45 ksi at room temperature) and where good weldability is desired. The temper designation T651 identifies the alloy condition as solution heat-treated and stress relieved by stretching.

Titanium alloys also have high strength and low density. The alloy used most often by Goddard is Ti-6Al-4V. The 6 and 4 indicate the weight percents of the aluminum and vanadium respectively.

This is an alpha-beta titanium alloy and was studied in the annealed condition. This alloy is much stronger than the aluminum alloys (tensile strength is 130 ksi at room temperature) but is more dense.

C17200 is a copper beryllium alloy which is used in applications such as springs and electrical contacts which require stiffness and/or good conductivity. This alloy is also heat-treatable and the designation TF00 identifies the temper as solution heat-treated and precipitation-hardened.

Fracture Tests

Two types of fracture tests were chosen to simulate instantaneous as well as fatigue failures. The tests were performed according to procedures outlined by the American Society for Testing and Materials (ASTM).

The fracture tests selected for the handbook include:
1. Standard tension test with a smooth gage area specimen.
2. Standard tension test with a notched gage area specimen.
3. Fracture toughness test with a compact tension specimen.

The tension tests were conducted at both room temperature and 77 K (Liquid nitrogen temperature). The fracture toughness tests were conducted at room temperature only. All room temperature tests were performed in air.

Format of Data Presentation

For each material, the nominal and actual composition, name of the supplier, heat treatment and processing, and metallographic information is provided. For each specimen at each condition for that alloy, a low magnification optical photograph of the fracture surface, optical photographs of a polished cross-section etched to reveal the microstructure, test data, and representative SEM fractographs are presented. Each alloy is identified by a separate identification number and each photo associated with that alloy has the alloy identification number as a prefix, i.e. 1-13 would be the 13th picture associated with the alloy identified by the number 1. The general material and processing information is located with the first test condition of a particular material.

Chemical Composition and Heat Treatment of the Alloys

A sample of each alloy was sectioned and analyzed to verify the chemical composition. The 6061 aluminum and Ti-6Al-4V were analyzed with the energy dispersive spectrometer (EDS); the 7075 aluminum and Cu-Be were analyzed using optical emission spectroscopy (OES).

Hardness tests and metallography were performed on each alloy to verify the heat treatment and grain structure. All the materials had the correct composition and heat treatment.

Preparation of Test Specimens

Two types of loading were chosen to simulate service failures: tensile overload and tension-tension fatigue.

Tensile Overload Specimens

For tensile overload testing, a machined specimen of the material is stressed to a load sufficient to cause rupture. This test would simulate a failure due to a one-time loading sufficient to cause fracture in service.

The specimens are prepared according to American Society for Testing and Materials (ASTM). Two types of test specimens were prepared; one with a smooth machined gauge area made in accordance to ASTM A370 (Figure 1), and the other with a circumferential notch in the machined gauge area made in accordance to ASTM E602-81 (Figure 2). The specimens are designed to have a reduced cross-section at mid-length for a uniform distribution of stress and to localize the zone of fracture. The smooth sample simulates a bulk alloy and the notched sample simulates a very sharp stress concentrator such as a fillet.

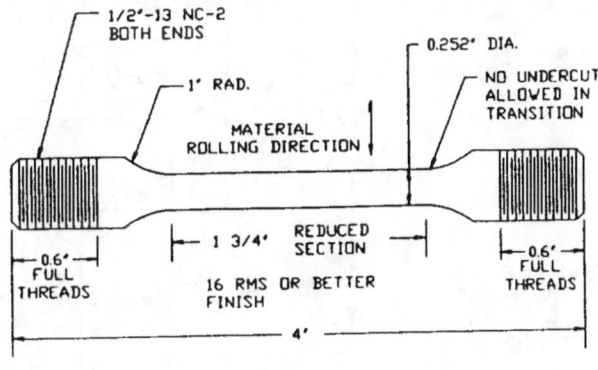

Figure 1. ASTM A370 Smooth Tensile Specimen

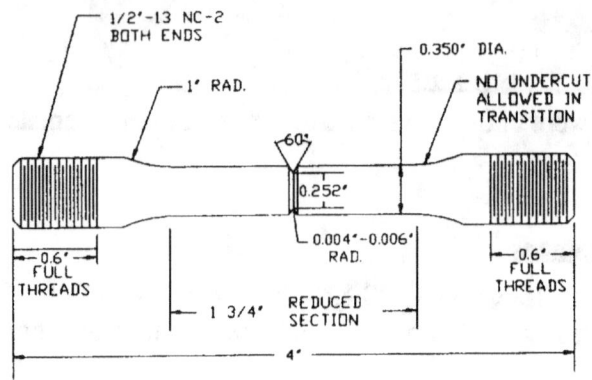

Figure 2. ASTM E602-1 Notched Tensile Specimen

Tension-Tension Fatigue Specimens

For tension-tension fatigue loading, a compact fracture toughness specimen is loaded cyclically until a preexisting crack grows. This test simulates a fatigue failure of an alloy in service. Fatigue crack growth in metals occurs due to repeated loading and unloading at the crack tip. Tension-tension fatigue specimens were manufactured according to ASTM 399 Appendix A4 (Figure 3).

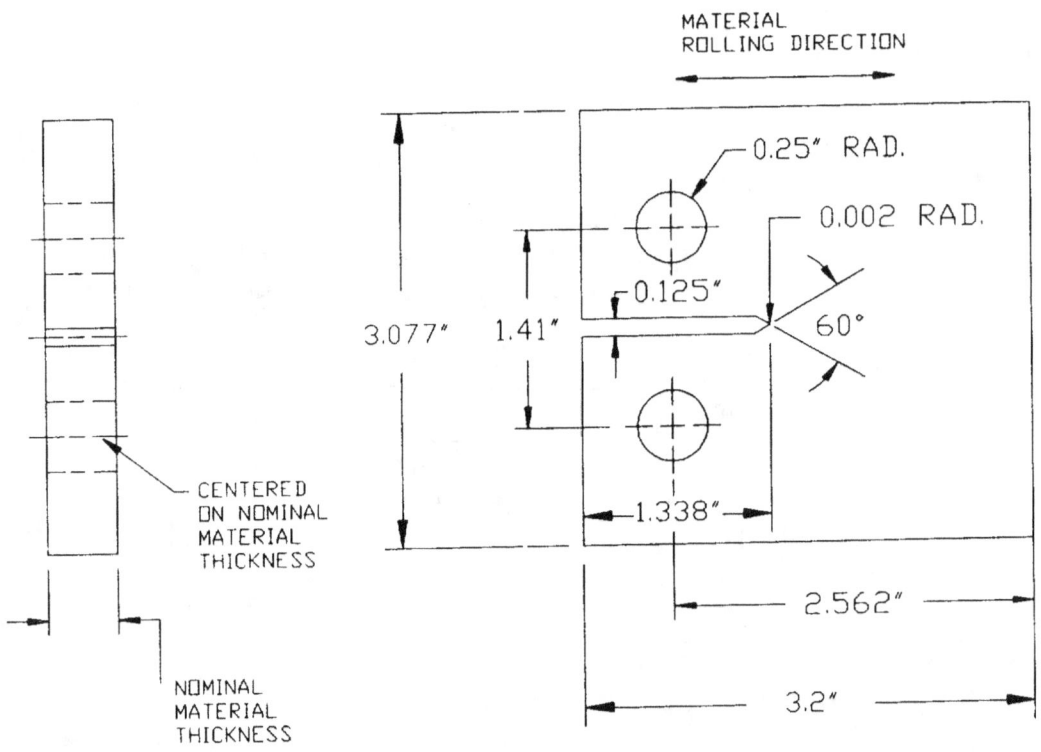

Figure 3. Compact tension specimen

ASTM 399 , the standard test for determining plane-strain fracture toughness (K_{Ic}) ,was used as a guide to produce fatigue crack growth in the alloy specimens. This standard was used to determine the loading required to produce fatigue crack growth.

In the ASTM test, the samples are cycled between a minimum and maximum load. For tension-tension fatigue all loading is tensile i.e, loads are positive. Tension-tension fatigue loading is chosen to allow the crack to remain open during testing thus preventing damage that might occur if the fracture surfaces were allowed to touch.

Two parameters are typically documented when reporting the ASTM test results; R and K_{max}. The R value is defined as the ratio of the minimum applied load to the maximum applied load:

$$R = min \ load/max \ load$$

If the R value is near 1.0, the difference between maximum and minimum loads is small. An R value near 0 indicates a large load difference.

The K value is the stress intensity at the crack tip and K_{max} is the maximum stress intensity at the crack tip. The larger the value, the higher the stress. K_{Ic} is the value of K where catastrophic failure occurs. Thus K is chosen to be less than K_{Ic} as to obtain crack growth without catastrophic failure.

Temperature

The materials were tensile tested at both room temperature and at 77 K (LN_2 or liquid nitrogen temperature) but were fatigue tested at room temperature only. Goddard experiments often operate at LN_2 temperature. The engineering models are often tested at 77 K to ensure the integrity of the spacecraft design and materials at low temperatures. If the fracture surface appears different for the same material at different temperatures, this handbook will help distinguish between a failure that occurred at room temperature or one that occurred at 77 K.

Appendix 1 is a chart summarizing the materials, test specimens, and temperatures we have used. A total of 40 tensile overload samples and 5 fatigue samples were tested and one sample at each condition was documented with fractographs.

Sample Preparation for Fractography

The first step in specimen preparation includes examining the specimen under a low-power optical microscope to document the initial conditions of the fracture. This step is required to identify features in case they are destroyed by cleaning and preparation techniques. Also an optical examination often identifies clues as to the cause of failure such as the fracture origin, overload features such as dimples, fatigue features such as fatigue rings etc.

The only requirement for clean samples is that they be conductive and small enough to fit in the specimen chamber. To meet these requirements, a portion of the fractured end is cut to approximately 1/2 inch in length and affixed to the specimen holder with two sided tape and conductive silver paint.

Contaminated samples are cleaned sequentially in baths of ethyl alcohol, acetone, and distilled water in an ultrasonic cleaner.

Fractography

Fractography is the documentation of fracture surfaces with photomicrographs. For this handbook, a Philips PSEM 500X scanning electron microscope (SEM) was used to make photomicrographs of the fractures at various magnifications. A SEM is superior to a light microscope for this work due to its better resolution (100 Å) and greater depth of field. The latter feature allows the high and low spots of a sample to be in focus in the same photograph.

Features encountered in tensile overload failure SEM fractography:

ductile dimples - cup-like features which are the direct result of a phenomenon called microvoid coalescence, which is the process by which ductile alloys fail when subjected to overload. Microvoids nucleate at regions of localized strain discontinuity such as that associated with second-phase particles, inclusions, grain boundaries, and dislocation pile-ups. As the strain increases, the microvoids grow, coalesce, and eventually form a continuous fracture surface.

equiaxed - The shape of the microvoid is governed by the state of stress within the material as the microvoids form and coalesce. Uniaxial tensile load, such as our tensile overload test, usually results in the formation of essentially *equiaxed* dimples (the shape being symmetric with respect of all axes). The microstructure and plasticity of the material determine whether the dimples will be deep and conical or shallow in shape.

elongated - Dimples that are elongated, that is stretched out in one direction, are associated with shear and tear dimples. The shear dimples have an oval shape with the long dimension pointing in the direction of shear. The elongated dimples on opposite faces of the fracture point in opposite directions. For tear failures, the dimples on the mating fracture faces point in the same direction. However, tear dimples are rare, which indicates that tensile tearing is not a predominant mechanism of fracture.

rolled/banded - words used to describe the highly oriented structure evident in some of the fracture surfaces of the worked alloys.

6

secondary cracking - Cracks appearing in the fracture surface either near the origin or throughout the surface of the fracture. The cracks can branch, they can have separate nuclei never crossing paths, and they can be transverse to the primary plane of fracture.

intergranular fracture - This type of fracture is simply described as grain boundary separation.

Features encountered in fatigue failure SEM fractography:

fatigue striations - Fracture that occurs as a the result of repetitive or fluctuating stresses is called fatigue fracture. There are three stages of fatigue fracture: initiation, propagation, and final rupture. We documented the features associated with the propagation stage.

The initiation stage is referred to as stage 1 fracture. Crack growth during this stage occurs principally by a slip-plane mechanism that causes irreversible changes in the crystal structure of the metal. This stage is usually confined to a few grains close to the fracture origin.

Stage 2 is the crack propagation stage where the fatigue crack tip progresses in microscopic advances causing a striated (striped) fracture morphology. These stripes are called *fatigue striations*. With each load cycle the crack forms one striation, given that the load is sufficient to propagate the crack. Striations are unique to fatigue failures (although not all materials exhibit fatigue striations) and they help distinguish the fracture from a tensile overload fracture. The *striation spacing* indicates the magnitude of the load applied to the metal and is therefore an important fractographic feature. The striations are only visible at high magnification and they can be very closely spaced.

The crack has reached stage 3 when the part is weakened to the point where one more load application will cause complete fracture. Stage 3 is called final rupture. The fracture mode may either be ductile, brittle or a combination of the two modes, depending on the material, the stress level, and the environment. For the alloys we tested, the materials behaved in a ductile manner.

Some of the features associated with the tensile overload fractures also appear on the Stage 3 fatigue fractures. These features include *secondary cracking*, *intergranular fracture* and *ductile dimples*.

Conclusion

This handbook was produced with intention of providing failure analysts a reference which can be consulted to compare fractures of known origin to those of a service failure. Another benefit is that NASA now has a fractography handbook which consists of materials and conditions often encountered in spacecraft applications.

Specimen 1   *7075-T76 Aluminum Wrought*

Composition:

(Nominal weight percents are maximum unless otherwise indicated).

Element	Nominal weight percent	Actual weight percent
Si	0.50	0.13
Fe	0.70	0.24
Cu	1.2-2.0	1.62
Mn	0.30	0.09
Cr	0.18-0.40	0.22
Zn	5.1-6.1	5.8
Mg	2.1-2.9	2.61
Ti	0.20	0.038

Supplier: Unknown (obtained from NASA/GSFC stock)

Specimen Condition:

Heat treatment: T76 - solution heat treated and stabilized. Hardness: HRB 83. Dimensions: 1"
Tk X 4.3" W X 14" L plate. The tensile test and fatigue specimens were cut and machined from
this plate as per Figure 1-a.

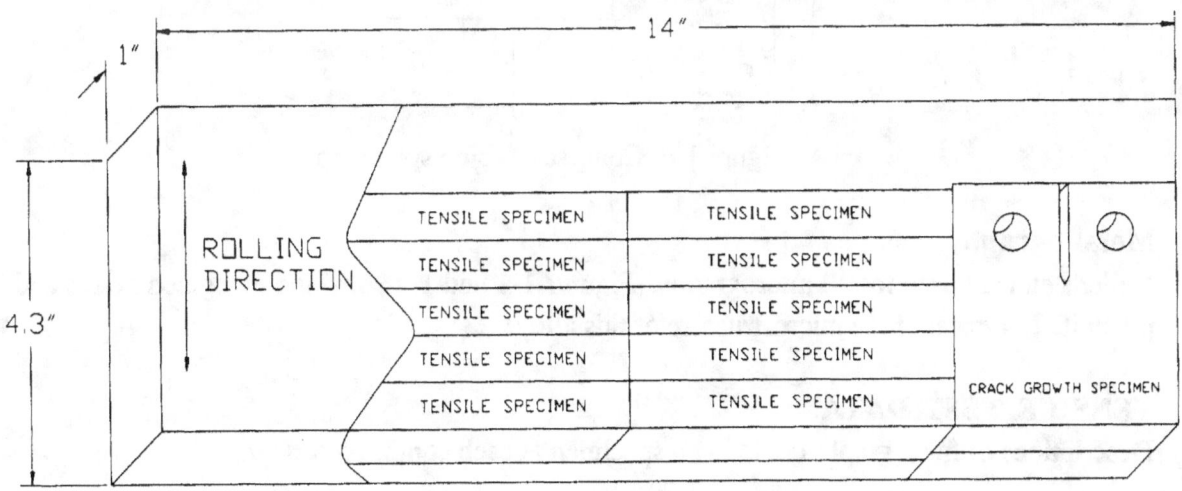

Figure 1-a. Pattern to cut test specimens

Specimen Geometry:

Smooth tensile specimens have dimensions as pictured in Figure 1-b. The notched tensile overload specimens have dimensions as pictured in Figure 1-c. The compact tension specimens were made according to the dimensions in Figure 1-d.

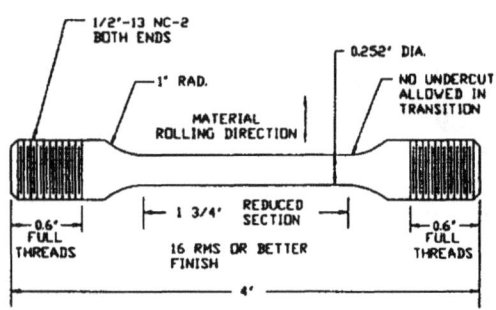

Figure 1-b. Smooth tensile overload specimen, 4" long, 0.252" diameter

Figure 1-c. Notched tensile overload specimen, 4" long, 0.004-0.006" root radius, 0.252" notch diameter.

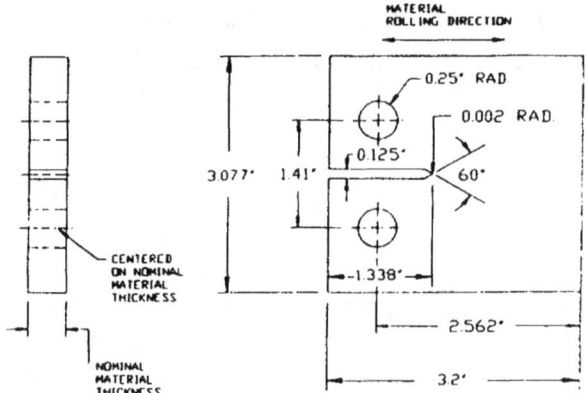

Figure 1-d. Compact tension specimen

Metallography:

Keller's etchant used for all micrographs. Figures 1-1 and 1-2 show the elongated grains and precipitates typical of the microstructure of this alloy.

TENSILE TEST DATA

Descriptions of fractographs depicting a specimen at each condition follow:

Smooth Tensile Overload at Room Temperature

Tensile Strength 79,000 psi Yield Strength 69,600 psi

Macroscopic Appearance of Fracture

The macrograph of this specimen is displayed in Figure 1-3. This specimen exhibits a cup-cone morphology with a rough surface and a pronounced shear lip. Two areas were documented; Area A which is near the center of the specimen; Area B which is located at the edge of the specimen where final fracture occurred by ductile shear.

Microscopic Appearance of Fracture

Figures 1-4 to 1-7 Area A: These micrographs show typical features of area A. The varied and elongated topography reflects the elongated grain structure of this alloy.

Figures 1-8 to 1-11 show typical features of area B. The shear lip has an oriented structure (Figures 1-8, 1-9) containing microvoids (1-10, 1-11). The features are similar to those in area A.

Smooth Tensile Overload at 77 K
Tensile Strength 96,500 psi

Macroscopic Appearance

The macrofractograph of this sample is shown in Figure 1-12. The surface of this fractured sample is rough and uneven and reflects the grain orientation that was produced by working the plate from the ingot. Area A represents the center of the sample and area B is near the edge of the sample.

Microscopic Appearance

Figures 1-13 to 1-15 Area A: Note the more brittle nature of the fracture than the room temperature case. We see very little evidence of ductile dimples except at high magnification as in Figure 1-15. There are many inclusions and step-like features.

Figures 1-16 to 1-18 Area B: These macrographs show the final shear lip. Here we see features similar to those in area A.

Notched Tensile Overload at Room Temperature
Tensile Strength 90,000 psi

Macroscopic Appearance

The surface looks rough yet planar due to the circumferential notch as seen in Figure 1-19. The planar surface is attributed to the notch acting as a stress concentrator and thus reducing plasticity. Rolling of the metal accounts for the oriented features. Two areas were chosen for close examination: Area A near the core of the sample; and area B which is close to the notched edge.

Microscopic Appearance

Figures 1-20 to 1-23 Area A: Ductility is evident in the form of dimples throughout the core of the sample. Inclusions and large dimples are seen in Figures 1-20, 1-21, 1-22. Note the steps in Figure 1-23.

Figures 1-24 and 1-25 Area B: The area near the notch has the same features as the core.

Notched Tensile Overload at 77 K
Tensile Strength 86,400 psi

Macroscopic Appearance

Figure 1-26 shows that the fracture surface looks much like that of its room temperature counterpart. The rolled, oriented structure is slightly more pronounced and the surface is flatter.

Microscopic Appearance

Figures 1-27 to 1-30 Area A: This area near the center of the specimen shows the same rolled appearance as the room temperature sample. There are fractured second phase inclusions in many of the elongated dimples. Note that very fine dimples are located on the ridges surrounding the inclusions in Figure 1-30.

Figures 1-31 to 1-34 Area B: These micrographs document an area closer to the notch. The rolled structure is accompanied by steps and transgranular cracks.

HIGH CYCLE FATIGUE TEST DATA
Specimen was fatigued at room temperature in tensile - tensile loading.

Macroscopic Appearance

Five distinct regions of fracture are visible in the SEM fractograph in Figure 1-35. Region 1 is the precrack region and region 5 is the fast fracture region. The other three regions were documented with SEM fractographs. These regions were loaded under the conditions listed in the chart below, where R represents the ratio of minimum to maximum load and K_{max} is the maximum crack tip stress intensity.

Region	R	K_{max}, (ksi-in$^{1/2}$)	da/dN
2	0.75	16.67	not available
3	0.50	16.67	not available
4	0.10	16.67	not available

Microscopic Appearance

Figures 1-36 to 1-38 Region 2: The striations are very faint and closely spaced.

Figures 1-39 to 1-41 Region 3: Several magnifications of the same area showing fairly largely spaced striations.

Figures 1-42 to 1-44 Region 4: Again the striation spacing appears fairly large which is expected for the small R value.

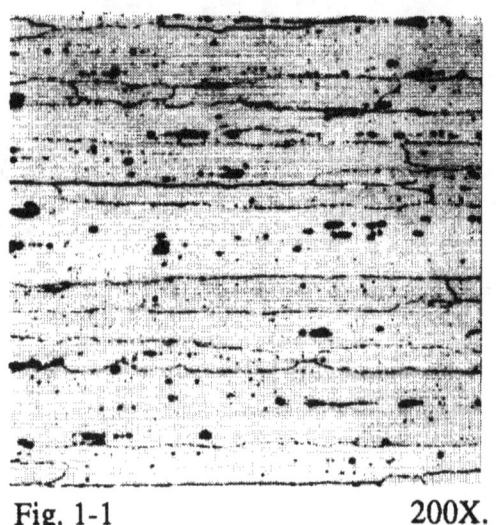

Fig. 1-1 200X.

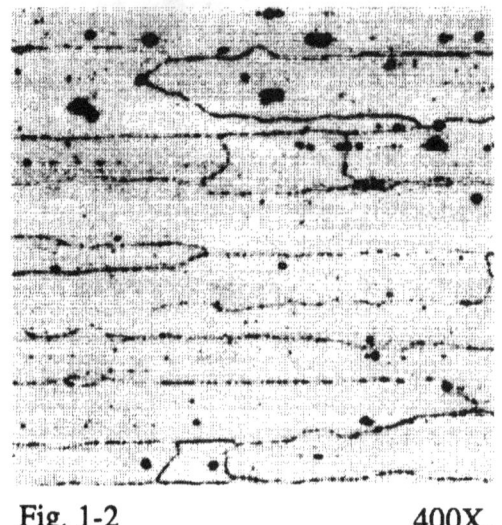

Fig. 1-2 400X.

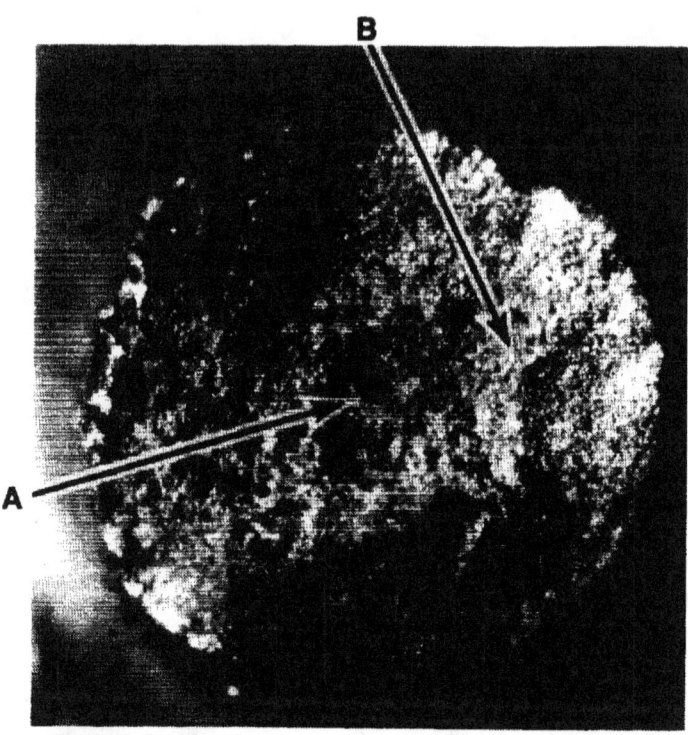

Fig. 1-3 15X
Optical photomacrograph of fracture surface.

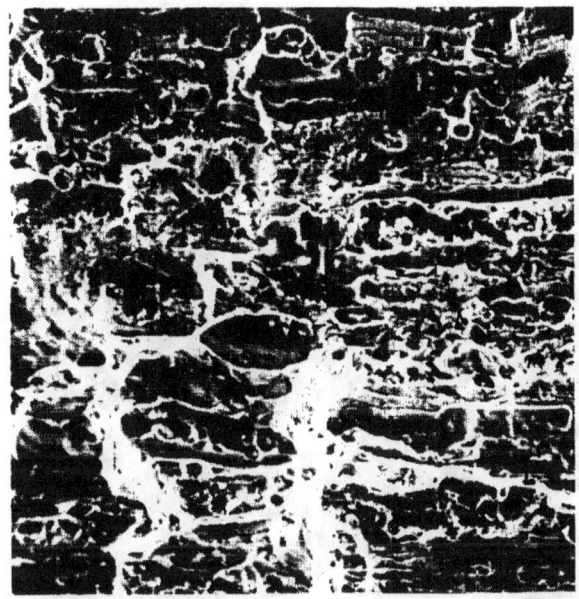

Fig. 1-4   160X
Region A. Area near the center of the
fracture suface.

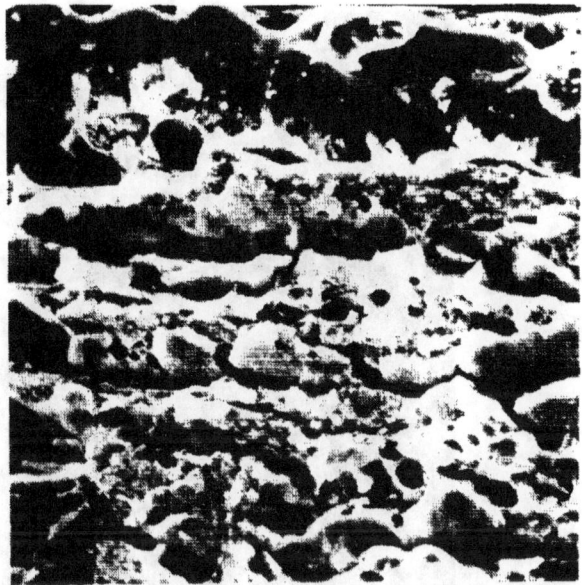

Fig. 1-5   640X
Region A. Enlargement of the center area
in Fig. 1-4.

Fig. 1-6   1250X
Region A. An area at the center of Fig. 1-5.

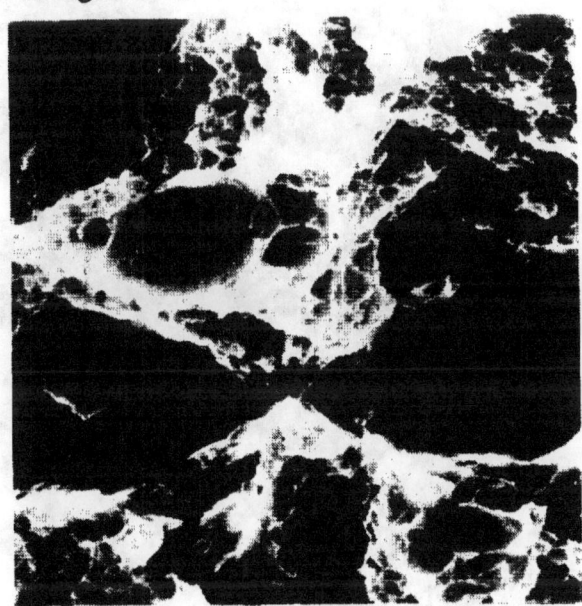

Fig. 1-7   5000X
Region A. An area at the center of Fig. 1-6.

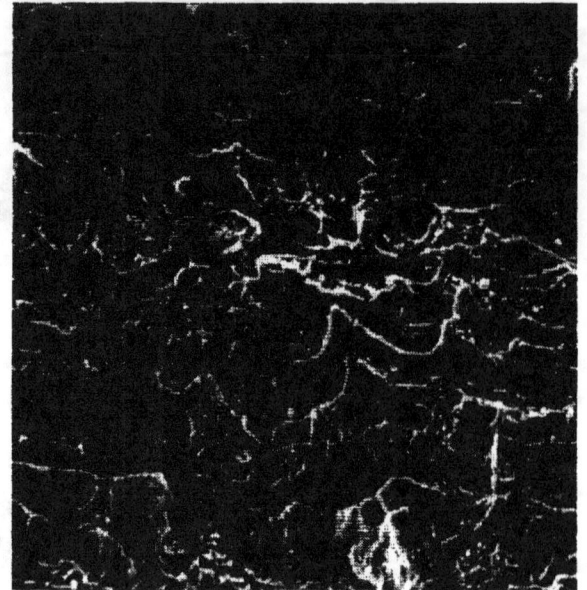

Fig. 1-8   160X.
Region B. An area on the shear lip.

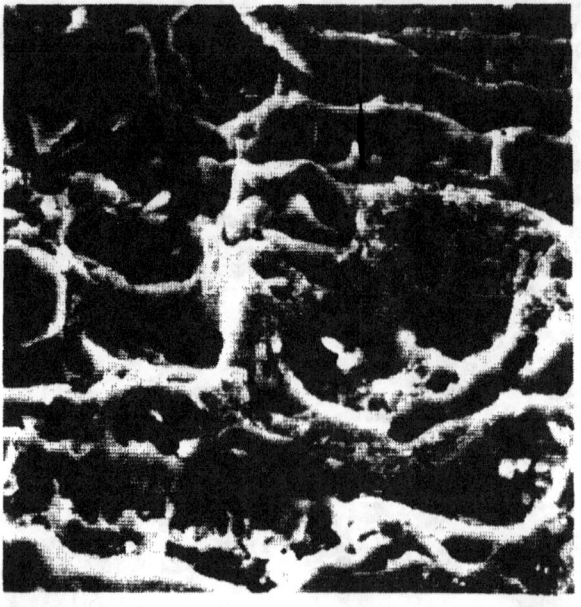

Fig. 1-9   640X
Region B. An area at the center of Fig. 1-8.

Fig. 1-10   1250X.
Region B. An area at the center of Fig. 1-9.

Fig. 1-11   2500X.
Region B. An area at the center of Fig. 1-10.

B

Fig. 1-12
Photomacrograph of fracture surface.
A
14X.

17

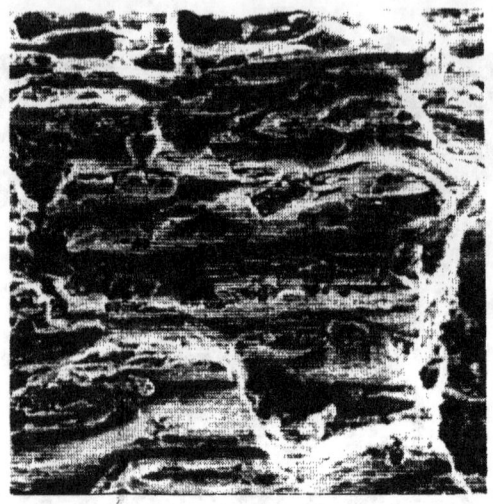

Fig. 1-13   160X
Area A. An area at the center of the
fracture.

Fig. 1-14   640X
Area A. An enlarged view of the center
area in Fig. 1-13.

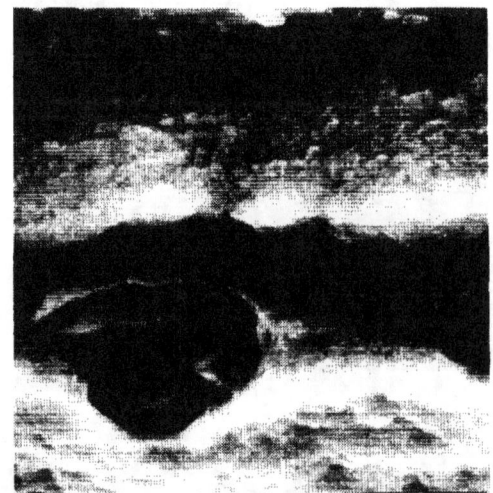

Fig. 1-15   2500X.
Area A. The brittle inclusions in
Fig. 1-14.

18

Fig. 1-16   80X
Area B. A low magnifcation picture
of the shear lip.

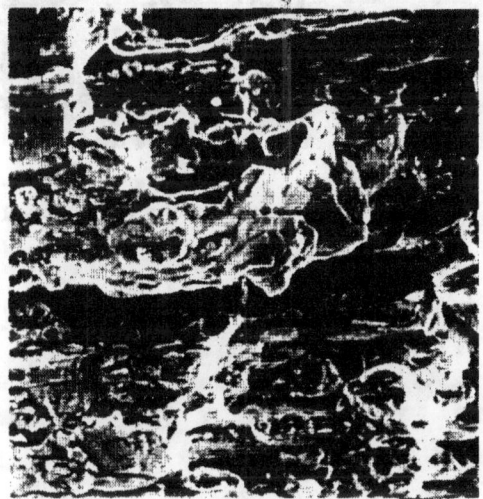

Fig. 1-17   160X
Area B. A closer view of the center
of Fig. 1-16.

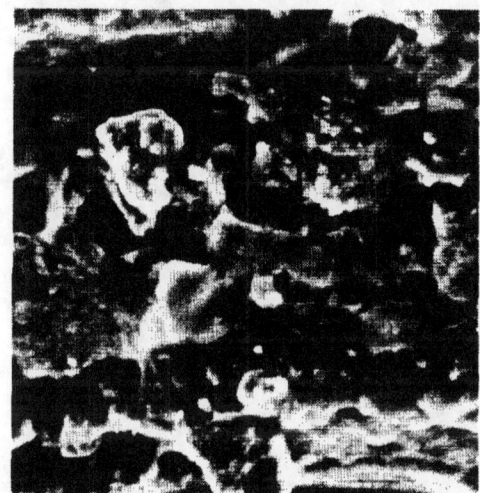

Fig. 1-18   640X
Area B. A closer view of the center
of Fig. 1-17.

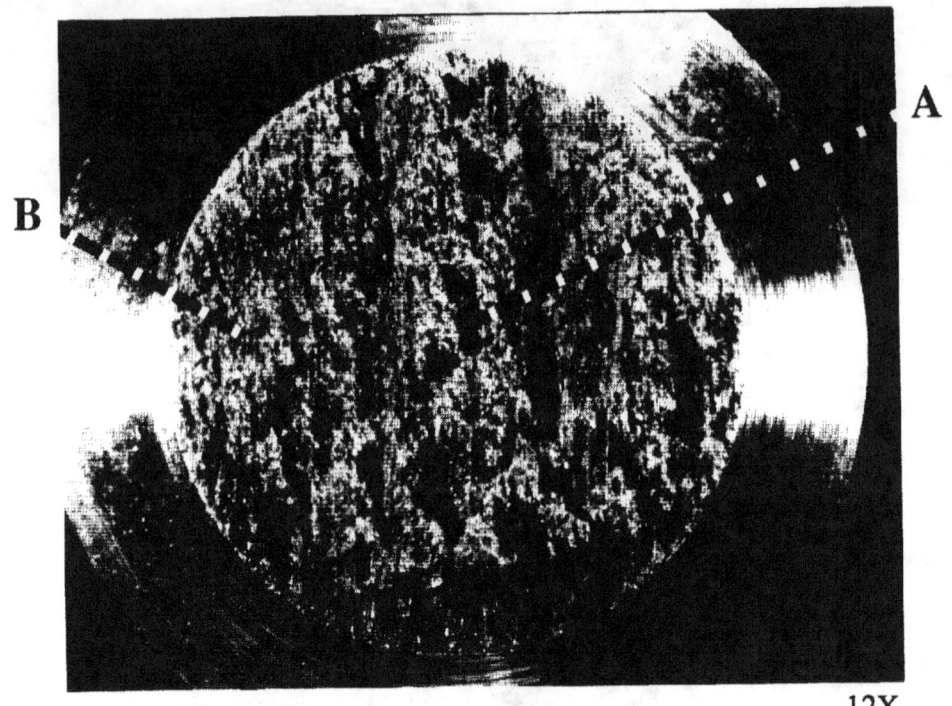

Fig. 1-19   12X
Optical photomacrograph of the fracture surface.

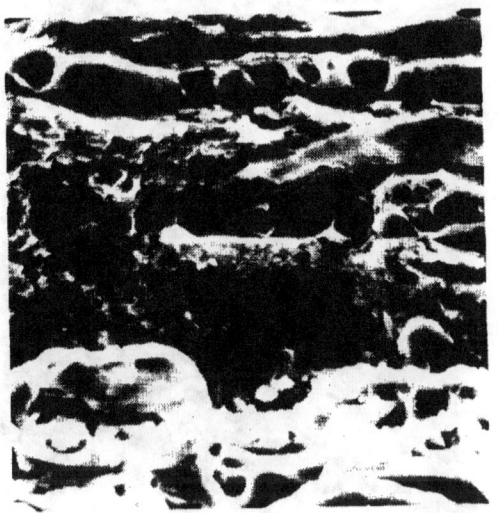

Fig. 1-20   640X
Area A. A view of the center of the
fracture surface.

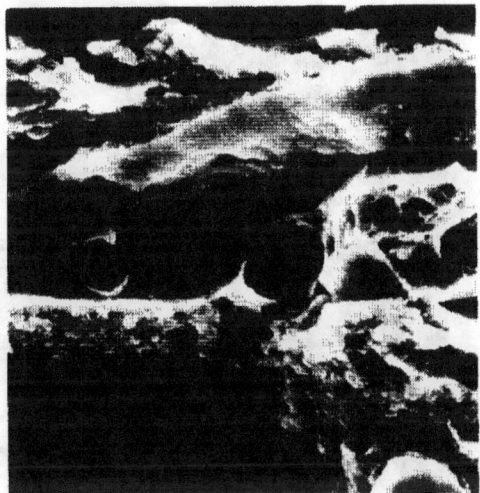

Fig. 1-21   1250X
Area A. A closer view of the center
of Fig. 1-20.

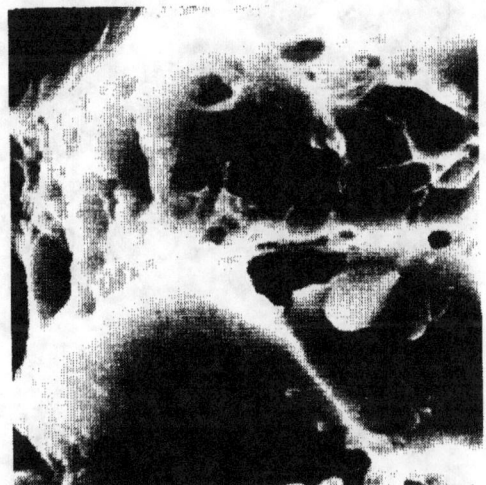

Fig. 1-22   5000X
Area A. Microdimples found in the
center of Fig. 1-21.

Fig. 1-23   160X
Area A. Steps located in the core of
the sample.

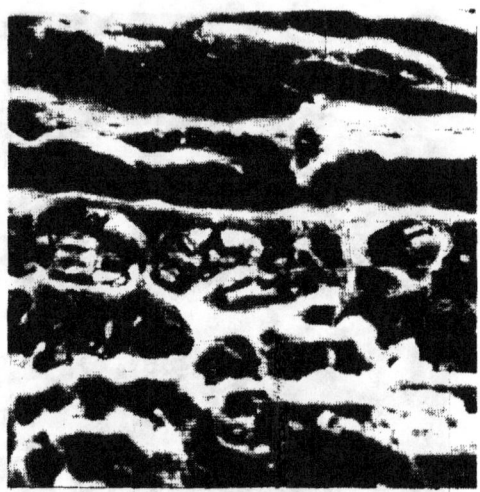

Fig. 1-24   640X
Area B. Similar features to those at
the core are found near the notch.

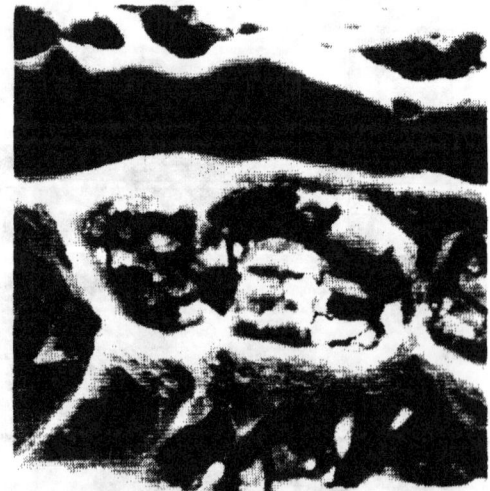

Fig. 1-25   1250X
Area B. A closer view of the center
of Fig. 1-24.

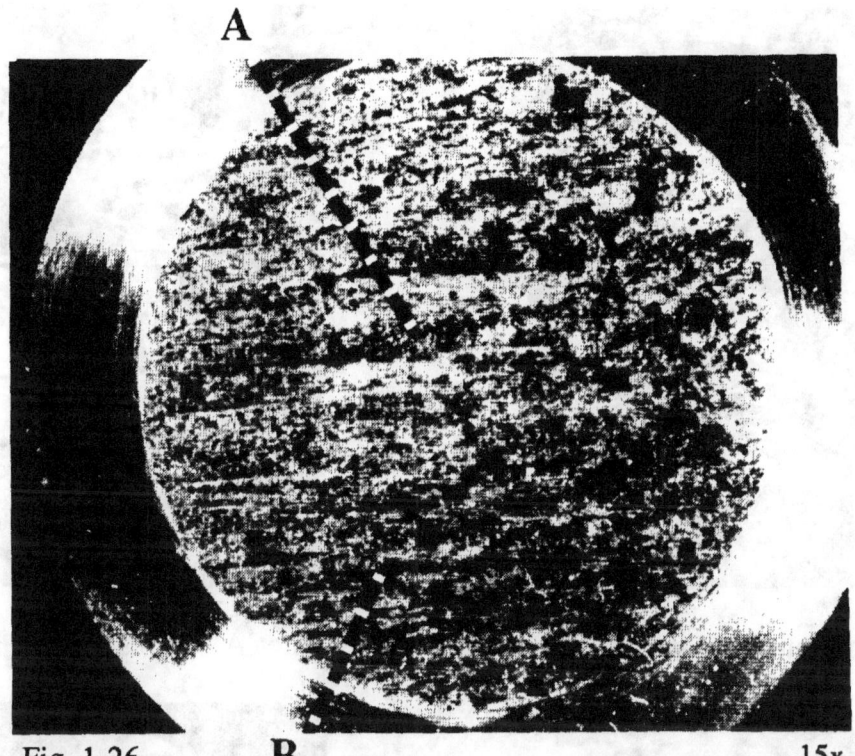

Fig. 1-26   15x
Optical photomacrograph of the fracture surface.

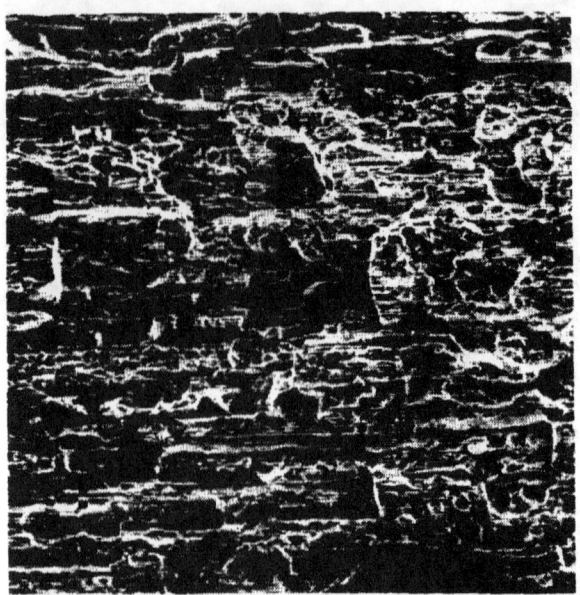

Fig. 1-27   80X
Area A. Rolled orientation and steplike
features.

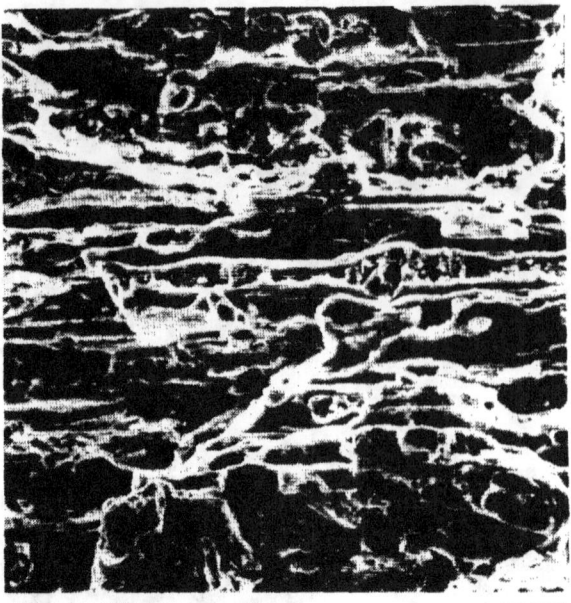

Fig. 1-28   320X
Area A. The center of Fig. 1-27.

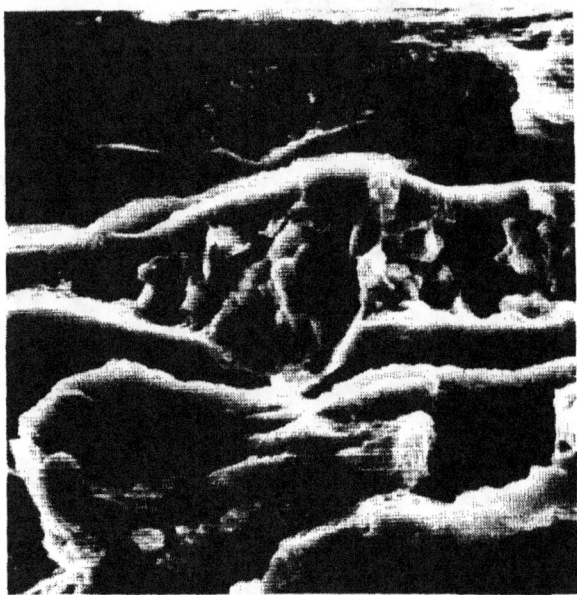

Fig. 1-29   1250X
Area A. An area in th center of Fig. 1-28.

Fig. 1-30   2500X
Area A. Fine dimples from Fig. 1-29.

Fig. 1-31   80X
Area B.

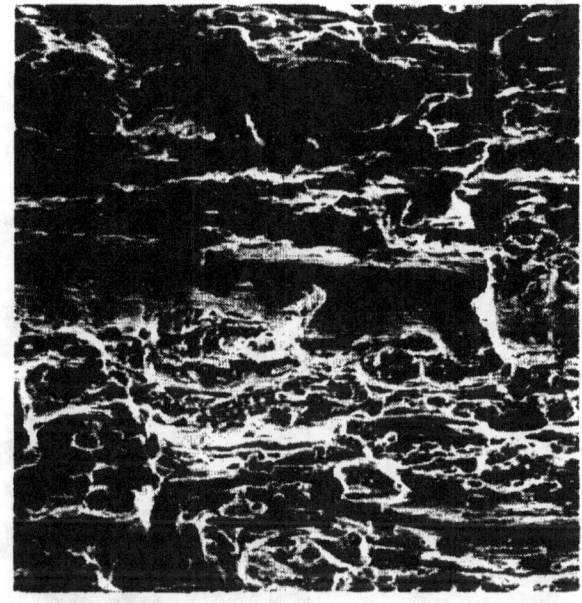

Fig. 1-32   160x
Area B. A closer view of Fig. 1-31.

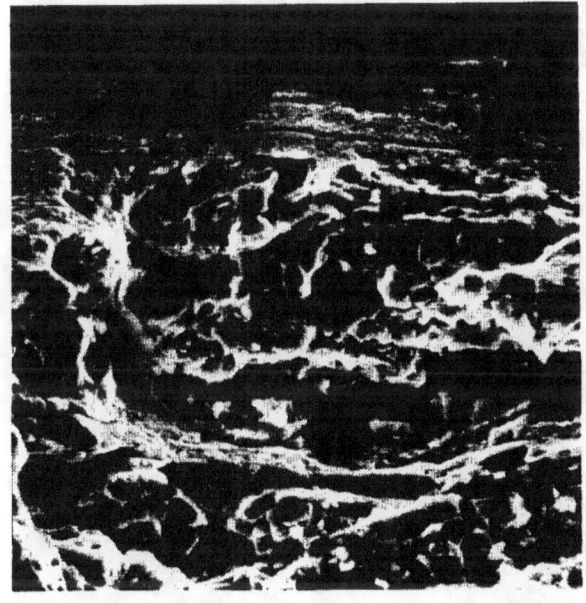

Fig. 1-33   640X
Area B.

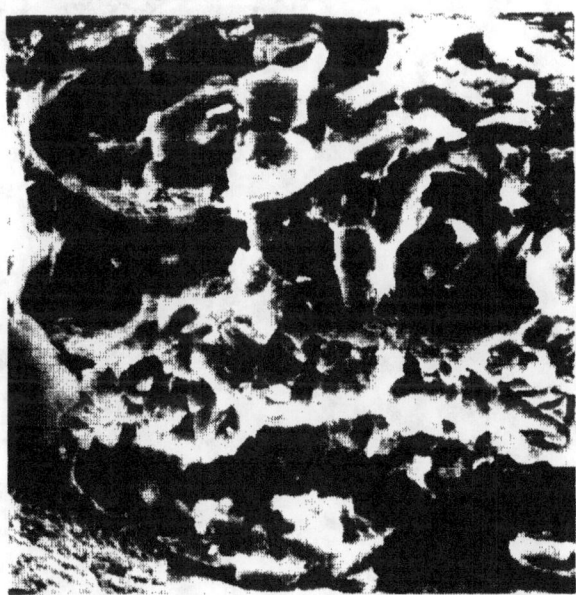

Fig. 1-34   1250X
Area B.

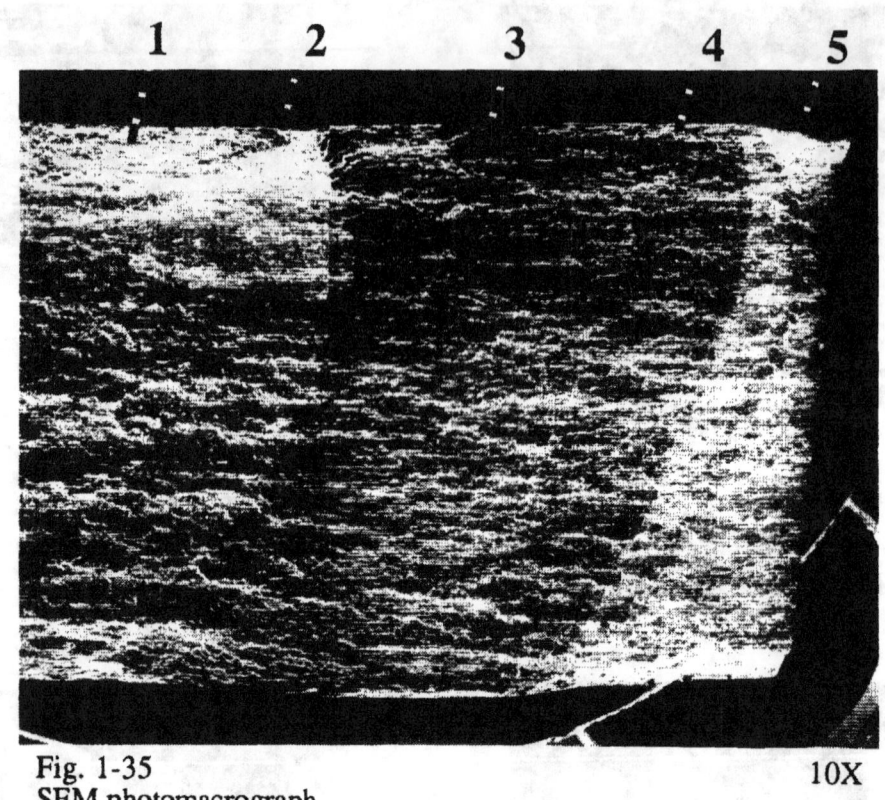

Fig. 1-35
SEM photomacrograph

10X

26

Fig. 1-36 1250X
Region 2. Faint fatigue striations.

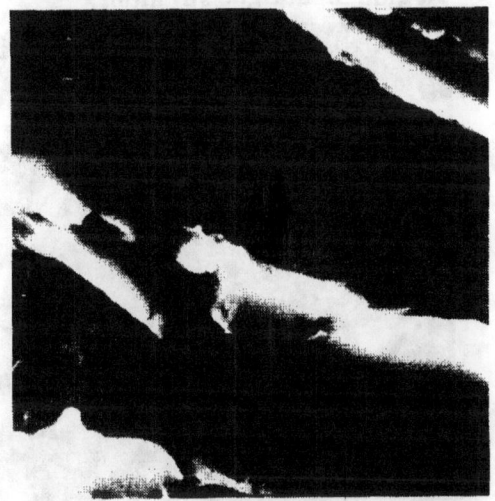

Fig. 1-37 2500X
Region 2.

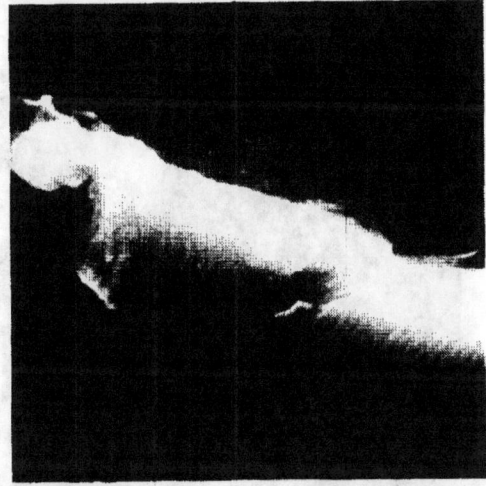

Fig 1-38 5000X
Region 2.

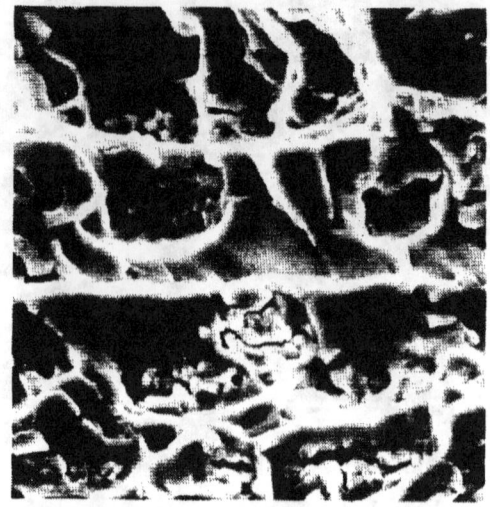

Fig 1-39   640X
Region 3. Fatigue striations.

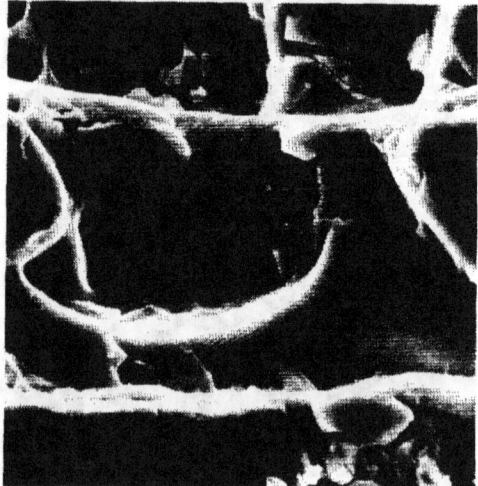

Fig. 1-40   1250X
Region 3.

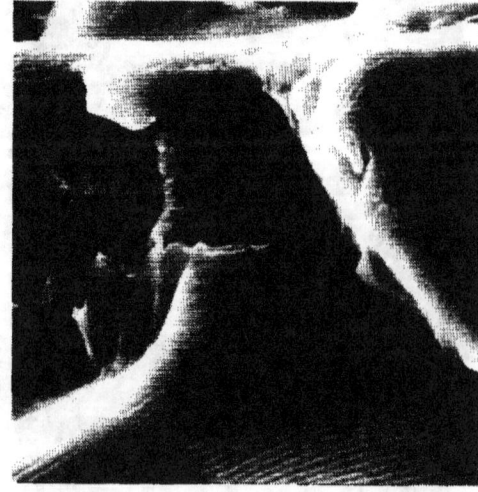

Fig. 1-41   2500X
Region 3

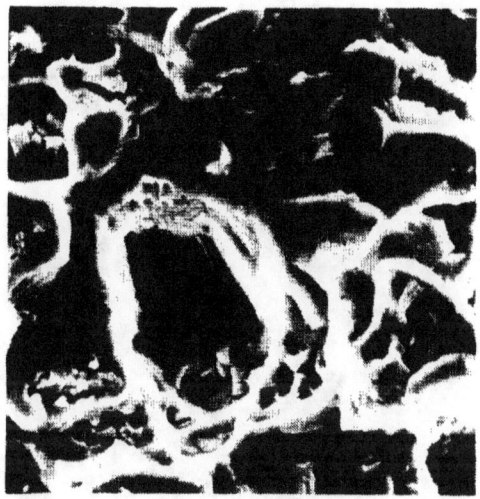

Fig. 1-42   640X
Region 4. Fatigue striations.

Fig. 1-43   1250X
Region 4.

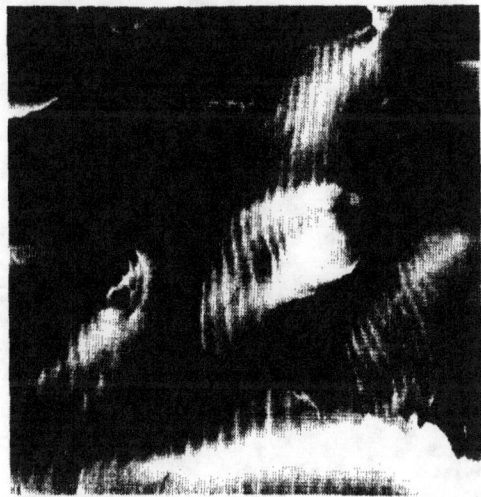

Fig. 1-44   2500X
Region 4.

Specimen 2   6061-T651 Wrought Aluminum

Composition:

Element	Nominal weight percent	Actual weight percent
Mg	0.8-1.2	1.02
Si	0.4-0.8	0.49
Cr	0.04-0.35	0.20
Mn	0.15 max	0.09
Fe	0.07	0.049
Cu	0.15-0.40	0.34
Zn	0.25 max	0.05

Supplier: Aluminum Company of America, Pittsburgh, PA (shipped by Metals Goods, Baltimore, MD)

Specimen Condition:
Initial condition was plate, mill finish, interleaved, per fed spec. QQ-A-230/11E. Heat treatment: T651- solution heat-treated and stress relieved by stretching. Hardness: HRB 58. Dimensions: 0.5" Tk X 46.5" W X 144.5" L. Specimens were cut from the plate according to the diagram in Figure 2-a.

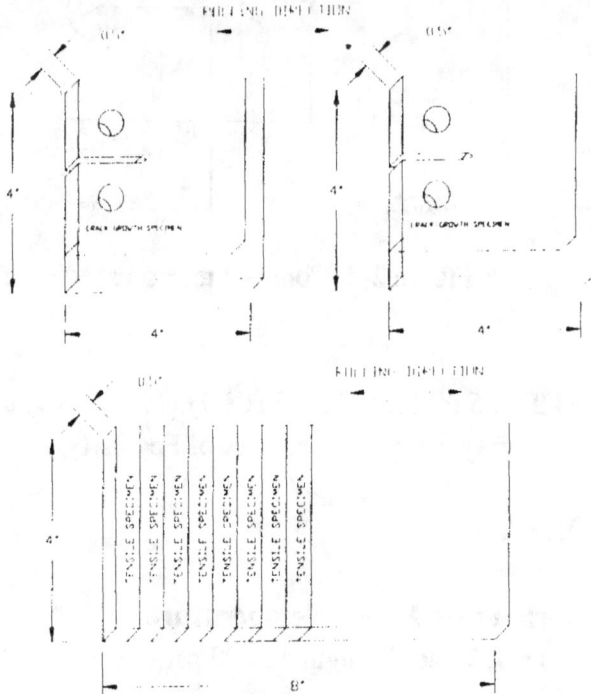

Figure 2-a, Pattern for cutting test specimens.

Specimen Geometry:

Smooth tensile overload specimens have the dimensions as shown in Figure 2-b and notched overload specimens have the dimensions as shown in Figure 2-c Compact tension specimens have the dimensions as shown in Figure 2-d.

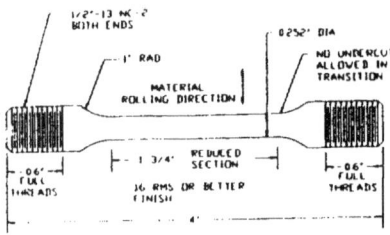

Figure 2-b. Smooth tensile overload specimen, 4" long, 0.252" diameter

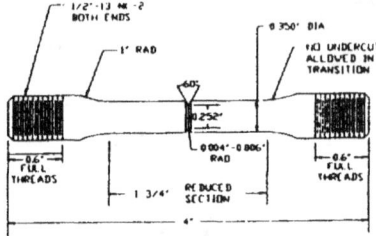

Figure 2-c. Notched tensile overload specimen, 4" long, 0.004-0.006" root radius, 0.252" notch diameter.

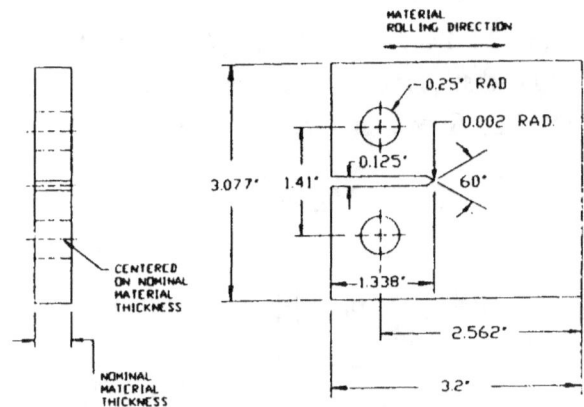

Figure 2-d, Compact tension specimen.

Metallography:

Keller's etchant (1 ml HF, 1.5 ml HCl, 2.5 ml HNO_3, 95 ml H_2O) used for all micrographs. Figures 2-1 and 2-2 show gray, scriptlike particles of Fe_3SiAl_{12}.

TENSILE TEST DATA

Smooth Tensile Overload at Room Temperature
Tensile Strength 47,200 psi, Yield Strength 41,800 psi

Macroscopic Appearance of Fracture

The fracture surface shown in Figure 2-3 has a central region which is composed of two ridges running parallel to two distinct shear lips displaying different inclinations to the specimen axis. Area A is of the shear lip and Area B is near the center of the sample.

Microscopic Appearance

Figures 2-4 to 2-6 Area A: These micrographs show typical features of area A on the upper and lower shear lip. Note the ductile dimples and the brittle inclusions configured in layers which reflect the oriented structure.

Figures 2-7 to 2-9 Area B: These macrographs show features typical of the center of the sample. The features are similar to those found on the shear lip.

Smooth Tensile Overload at 77 K
Tensile Strength 59,300 psi

Macroscopic Appearance

The macrograph of this smooth 77 K temperature tensile overload sample is displayed in Figure 2-10. This specimen exhibits a rough central region with a shear lip and jagged point on the edges. Area A represents the shear lip and B the central zone of fracture.

Microscopic Appearance

Figure 2-11 to 2-13 Area A: This area represents the shear lip region. Note that the dimples are smaller than in the smooth room temperature sample. Also the layered orientation is less evident. The colder temperature has impeded necking and plastic deformation.

Figures 2-14 and 2-16 Area B: The dimples appear larger in the central zone of the fracture than on the shear lip. There are more features such as steps and small ridges here.

Tensile Overload Notched Sample at Room Temperature
Tensile Strength 70,900 psi

Macroscopic Appearance

The overall appearance of this fracture surface as seen in Figure 2-17 is coarse and grainy. Area A depicts an area near the edge and area B depicts an area in the central region.

Microscopic Appearance

Figures 2-18 to 2-20 Area A: There are numerous ductile dimples. Note the very small dimples at 2500X in Figure 2-20.

Figures 2-21 to 2-23 Area B: This area looks similar to A but notice the shape of the ductile ridges. The unusual shape is due to parallel rows of inclusions. Microductility is also seen here.

Tensile Overload Notched Sample at 77 K
Tensile Strength 85,000 psi

Macroscopic Appearance
This failure surface is rough and grainy as shown in Figure 2-24. Several areas in the central fracture zone were documented.

Microscopic Appearance
Figures 2-25 to 2-30: Fractographs from various areas show the similarity of this sample to the notched sample at room temperature. The cold temperature did not change the appearance of the fracture.

HIGH CYCLE FATIGUE TEST DATA

Macroscopic Appearance
Five distinct regions are identified in the optical photograph in Figure 2-31. Region 1 is the pre-crack region and region 5 is the fast fracture region. Regions 2, 3 and 4 were documented using the SEM fractographs. The loading data for these regions is displayed in the following chart.

Region	R	$K_{max.}$ (ksi-in$^{1/2}$)	da/dN
2	0.75	16.67	not available
3	0.5	16.67	not available
4	0.10	16.67	not available

Microscopic Appearance
Figures 2-32 to 2-34 Region 2: The striations are faint at this low load ratio.

Figures 2-35 to 2-37 Region 3: The fatigue striations are much more defined at this load.

Figures 2-38 to 2-40 Region 4: Note that the striation spacing is even larger at the smallest R value.

Fig. 2-1   200X

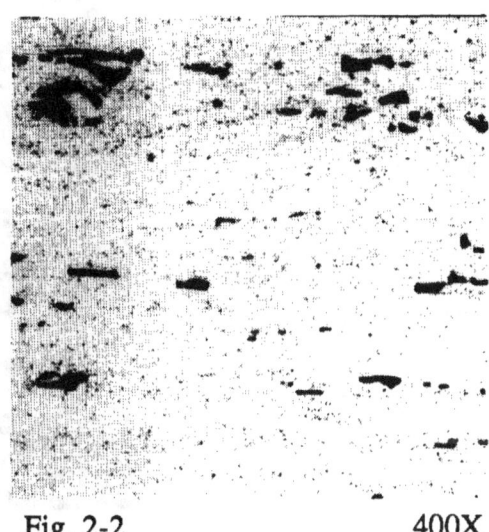

Fig. 2-2   400X

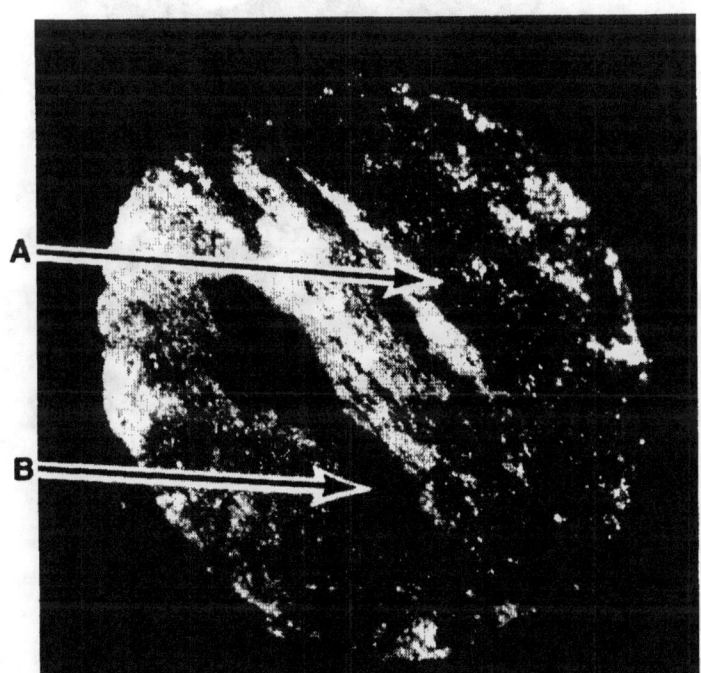

Fig. 2-3   15X
Optical photomacrograph of fracture surface.

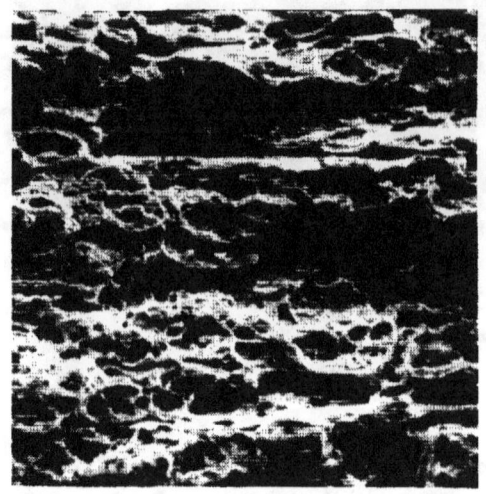

Fig 2-4   320X
Region A. Shear lip area.

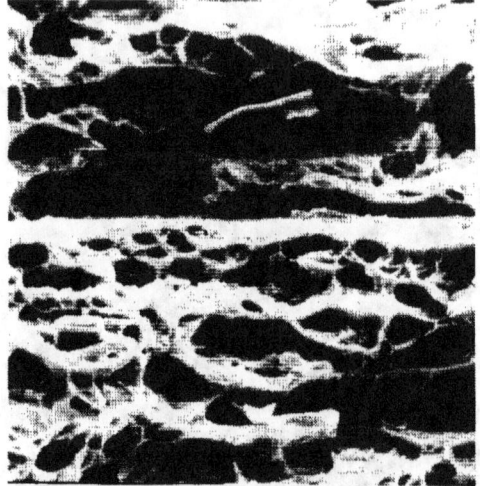

Fig 2-5   640X
Region A.

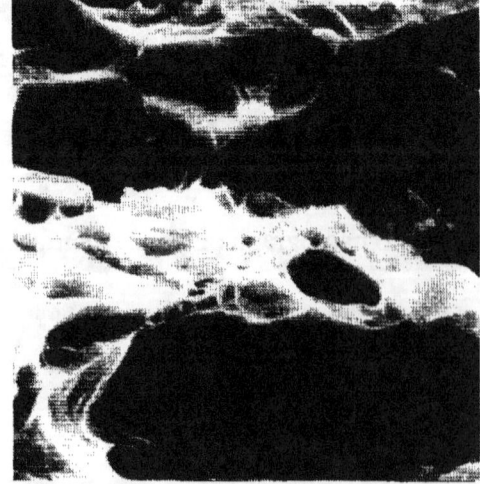

Fig 2-6   2500X
Region A. Ductile dimples.

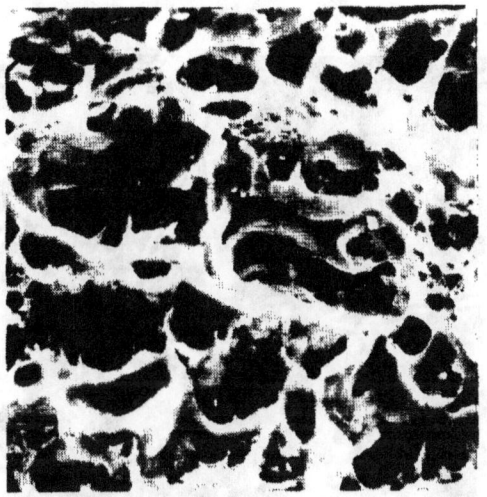

Fig 2-7   640X
Region B. Core of sample.

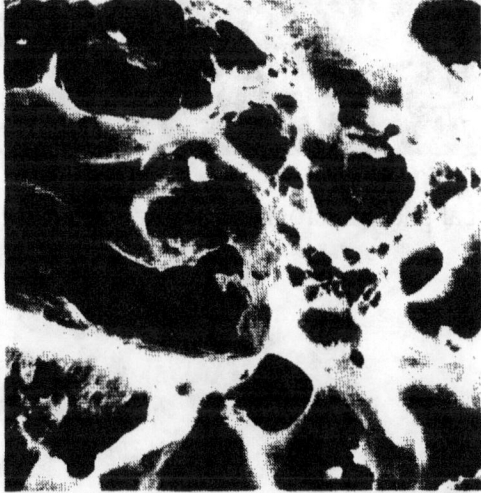

Fig 2-8   1250X
Region B. Large and small dimples.

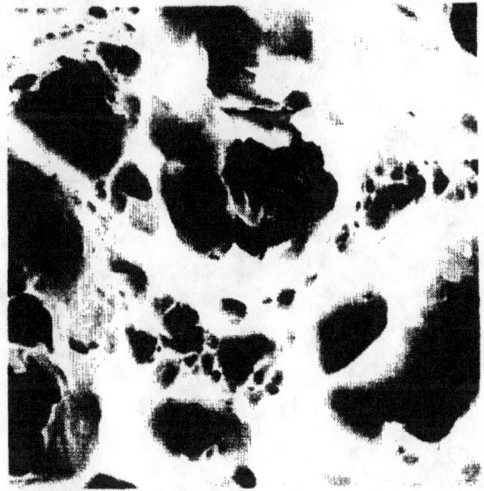

Fig 2-9   2500X
Region B. Microdimples.

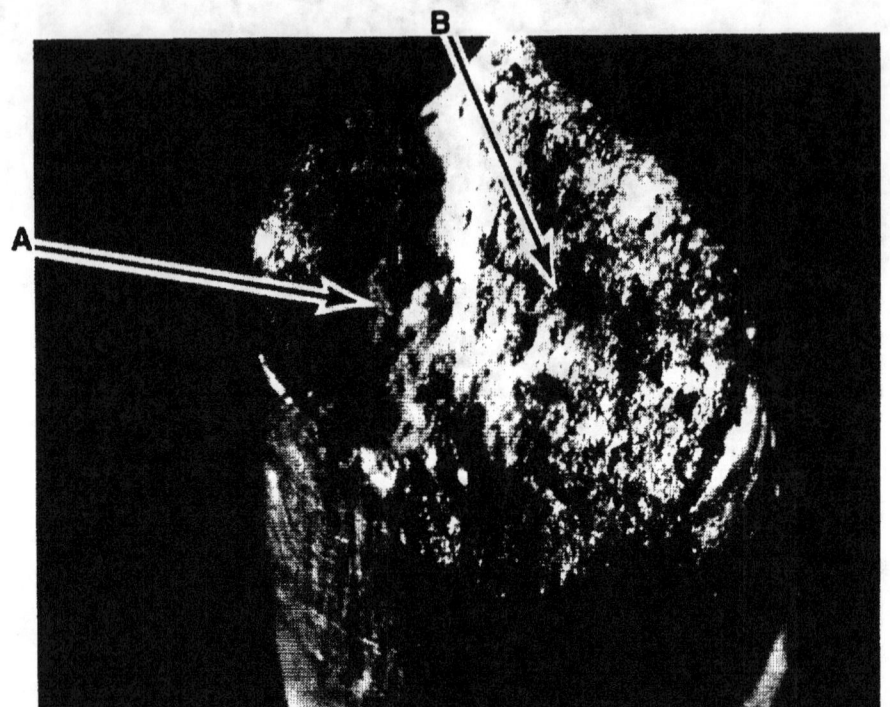

Fig. 2-10   15X
Optical photomacrograph of fracture suface.

38

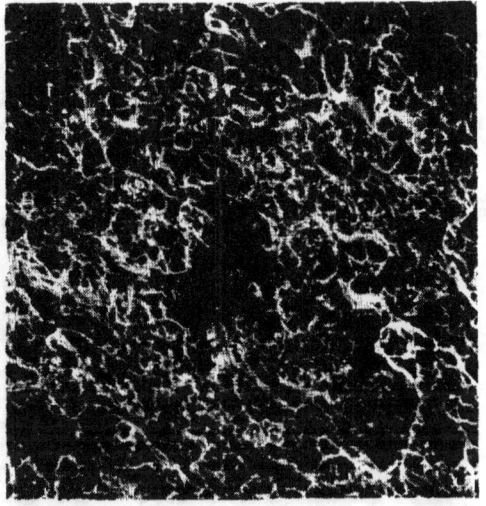

Fig. 2-11   160X
Region A. Shear lip region.

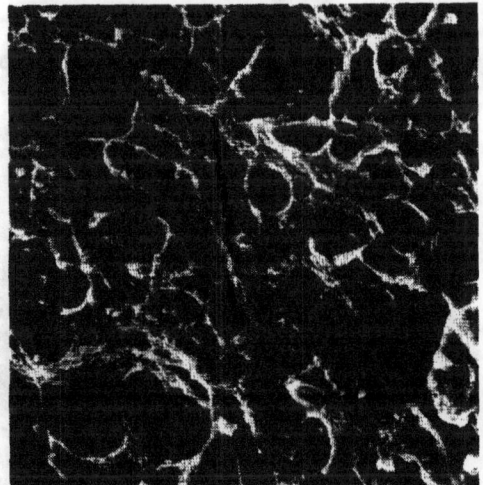

Fig 2-12   640X
Region A. Shear lip region. Note the
size variation of the ductile dimples.

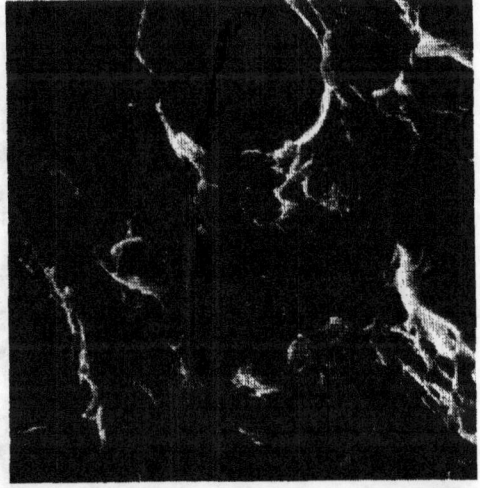

Fig 2-13   2500X
Region A. Microdimples in the shear
lip region.

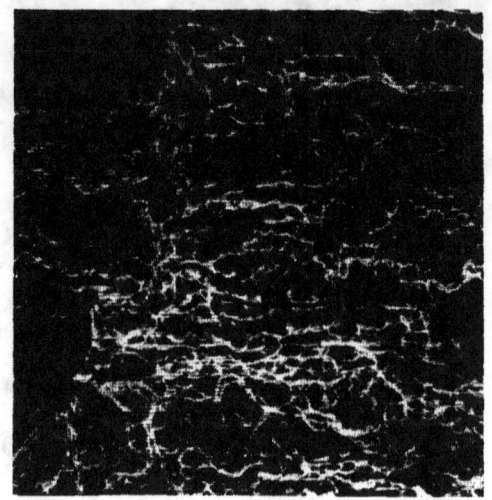

Fig 2-14   160X
Region B. Central fracture zone.

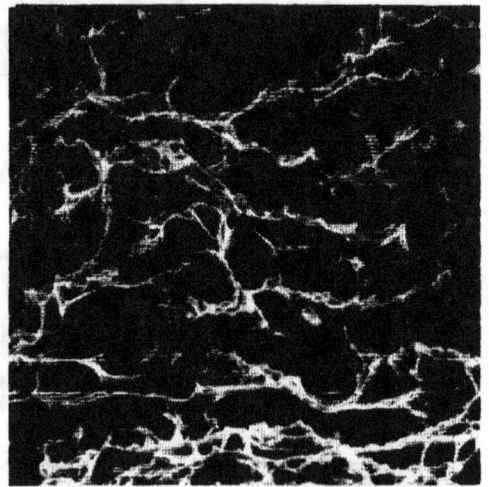

Fig 2-15   640X
Region B. The central area of
Fig. 2-14.

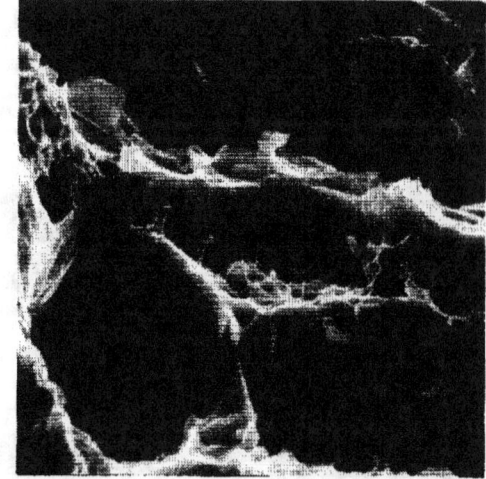

Fig 2-16   2500X.
Region B. Small ductile dimples.

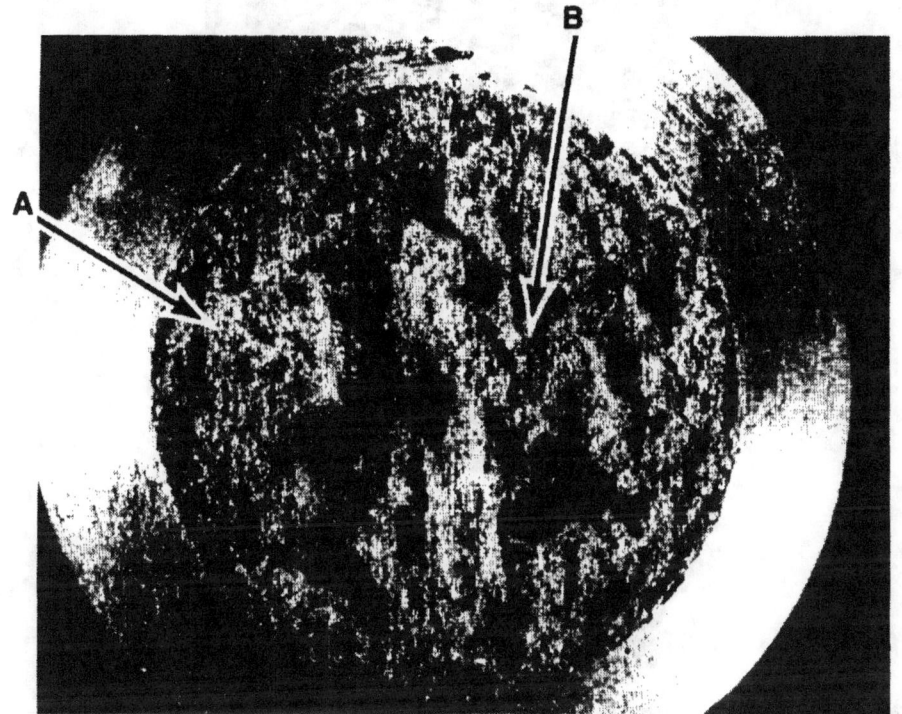

Fig 2-17
Photomacrograph.

13X.

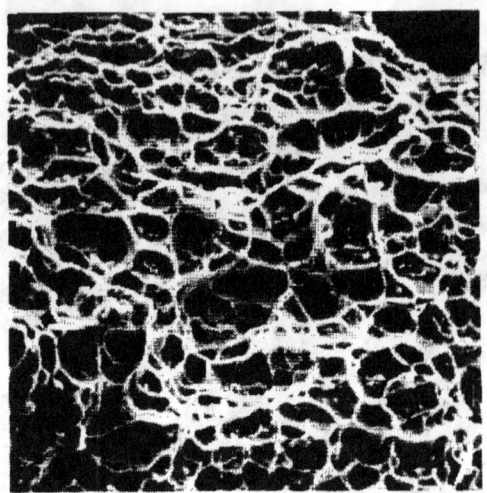

Fig 2-18   320X.
Region A. Central region of fracture.

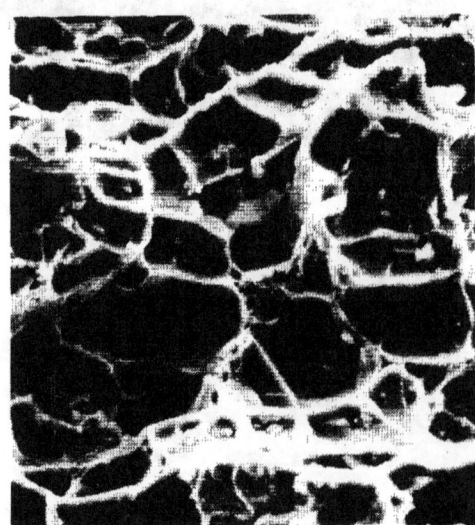

Fig 2-19   640X.
Region A. Unusual shape of dimples
 in central region.

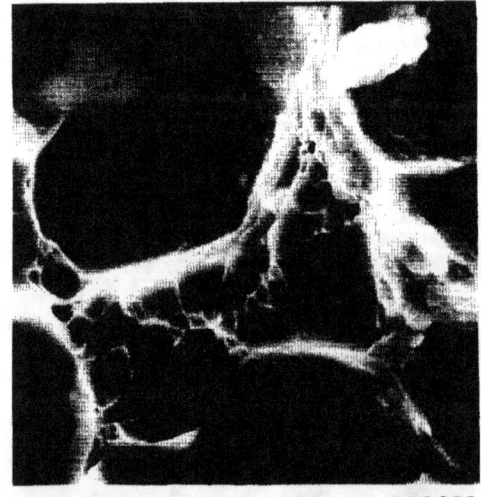

Fig 2-20   2500X.
Region A. Lower center of Fig. 2-19.

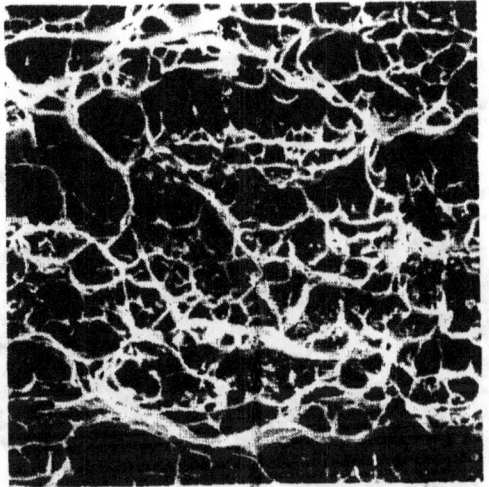

Fig 2-21   320X.
Region B. Area near the notch.

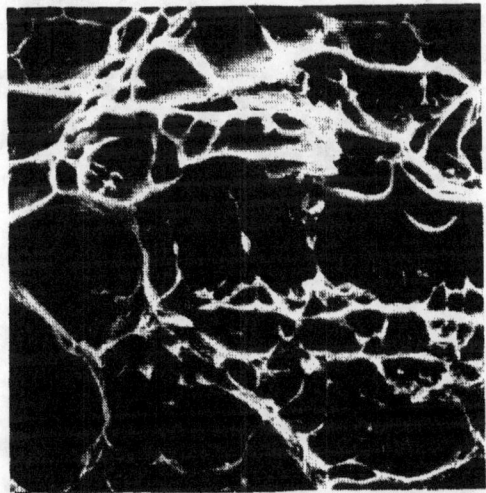

Fig 2-22   640X
Region B. Enlarged view of Fig. 2-21.

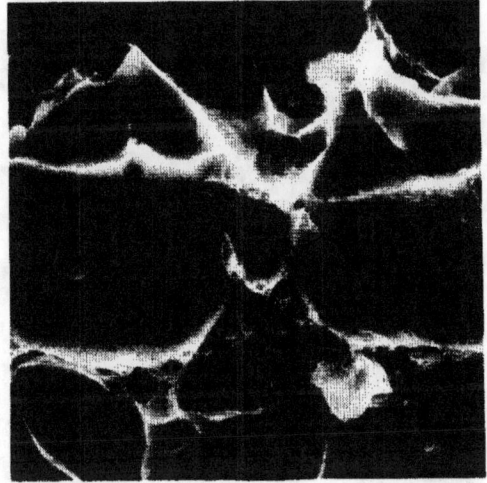

Fig 2-23   2500X
Region B. Microdimples in the shear
lip zone.

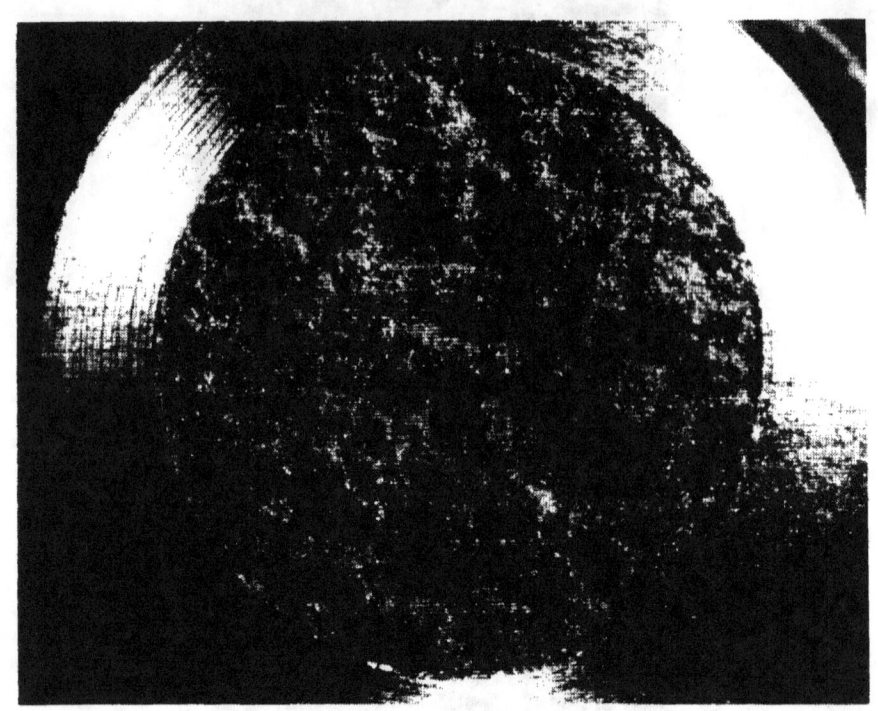

Fig 2-24   14X
Photomacrograph.

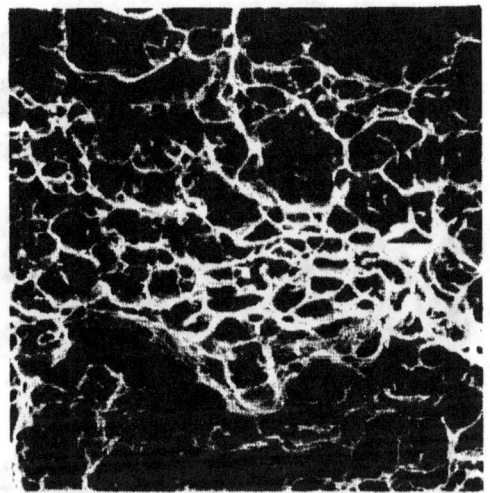

Fig 2-25   320X
Region A. Central region.

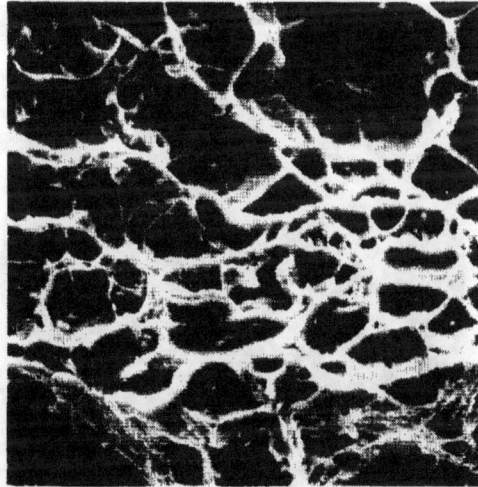

Fig 2-26   640X
Region A. Central region.

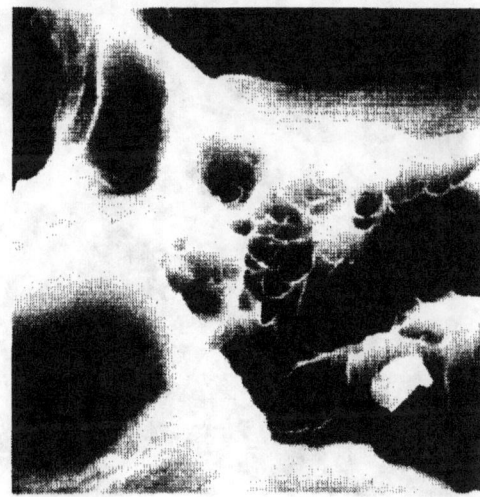

Fig 2-27   5000X
Region A. Central region.

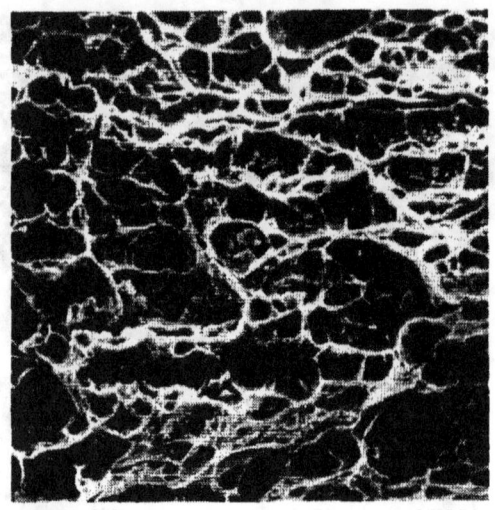

Fig 2-28   320X
Region A. Another area near the center.

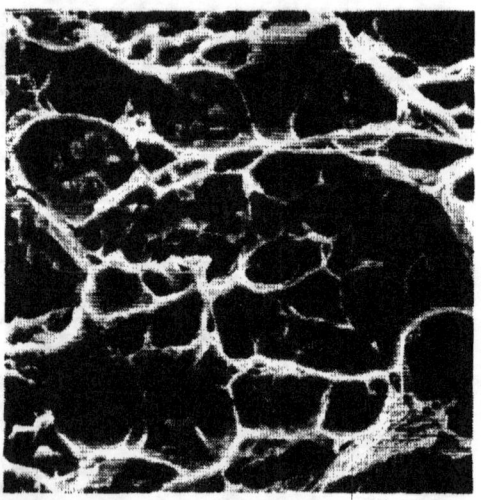

Fig 2-29   640X.
Region A. Same area as in Fig. 2-28.

Fig 2-30   2500X.
Region A. Same area as in Fig. 2-29.

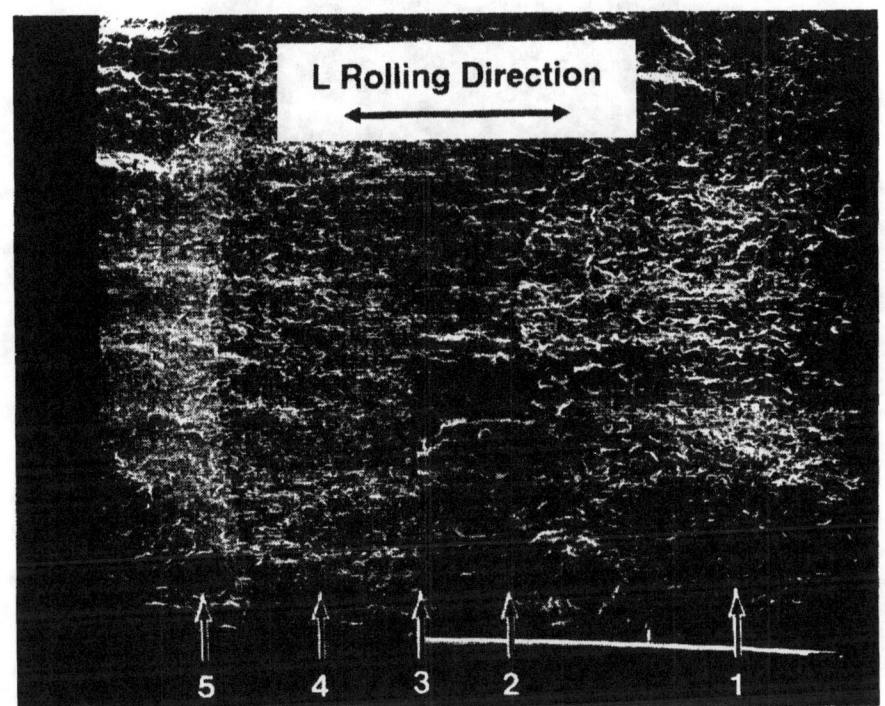

Fig. 2-31   10X
SEM photomacrograph of fracture surface.

47

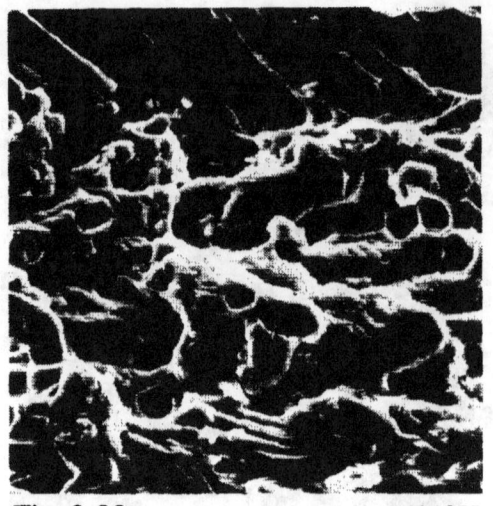

Fig. 2-32   640X
Region 2. Faint fatigue striations.

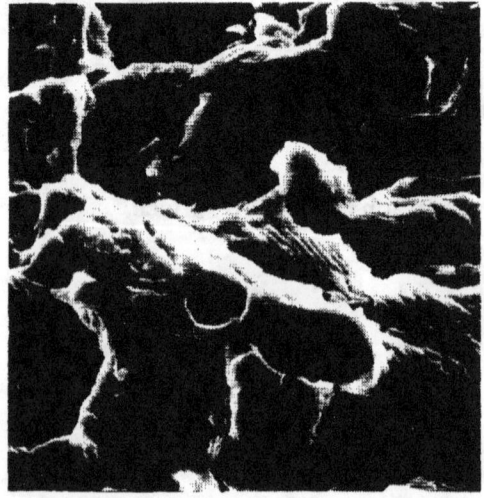

Fig. 2-33   1250X
Region 2.

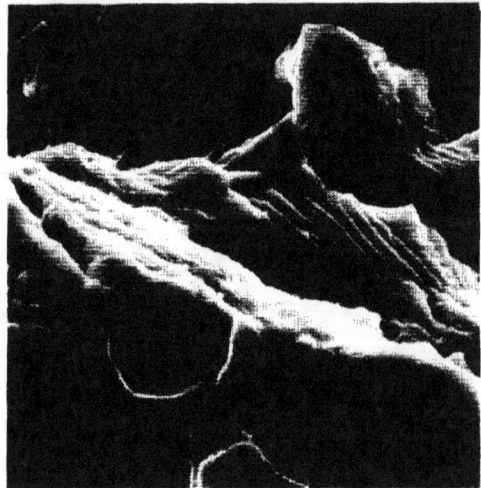

Fig. 2-34   2500X
Region 2.

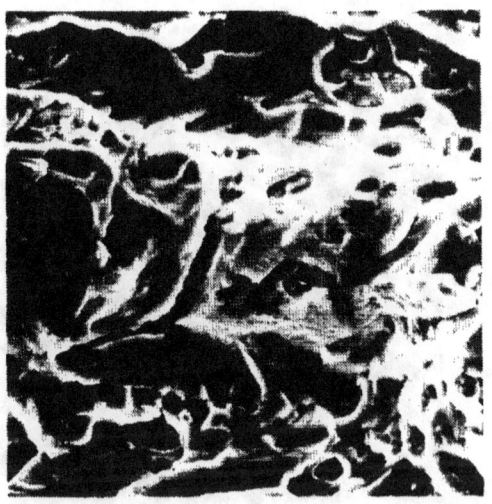

Fig. 2-35   640X
Region 3. Fatigue striations.

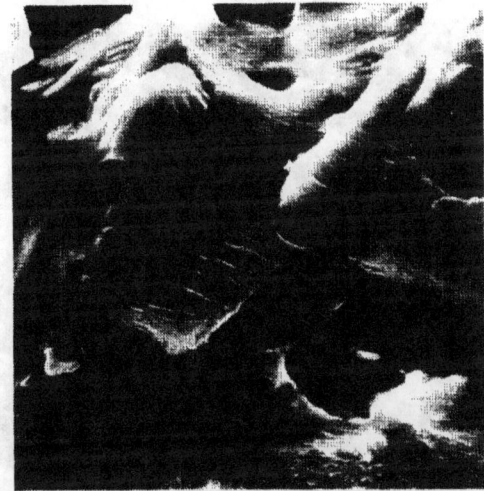

Fig. 2-36   2500X
Region 3.

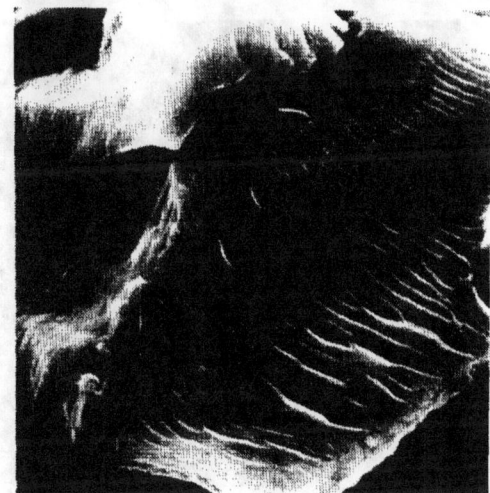

Fig. 2-37   5000X
Region 3.

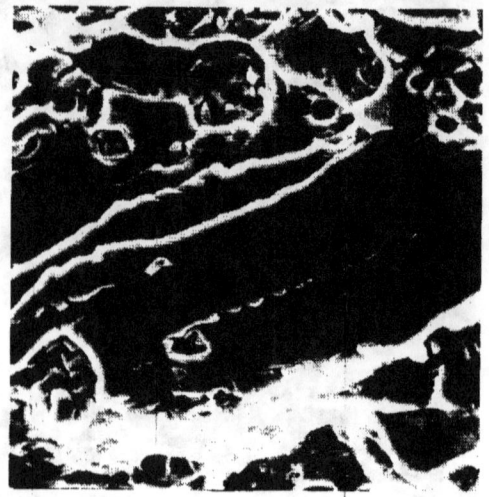

Fig. 2-38   640X
Region 4. Fatigue striations.

Fig. 2-39   1250X
Region 5.

Fig. 2-40   2500X
Region 5.

Specimen 3   Ti-6AL-4V Wrought

Composition:

Element	Nominal weight percent	Actual weight percent
Al	5.5-6.75	6.0
Ti	Balance	Balance
V	3.5-4.5	3.8

Supplier: RMI Company, Niles, Ohio 44446

Specimen Condition:

Mill annealed ELI plate. Heat treatment: As received annealed. Hardness: HRC 32. Dimensions: 0.434" Tk X 5" W X 18" L. Test specimens were cut from the plate according to the diagram in Figure 3-a.

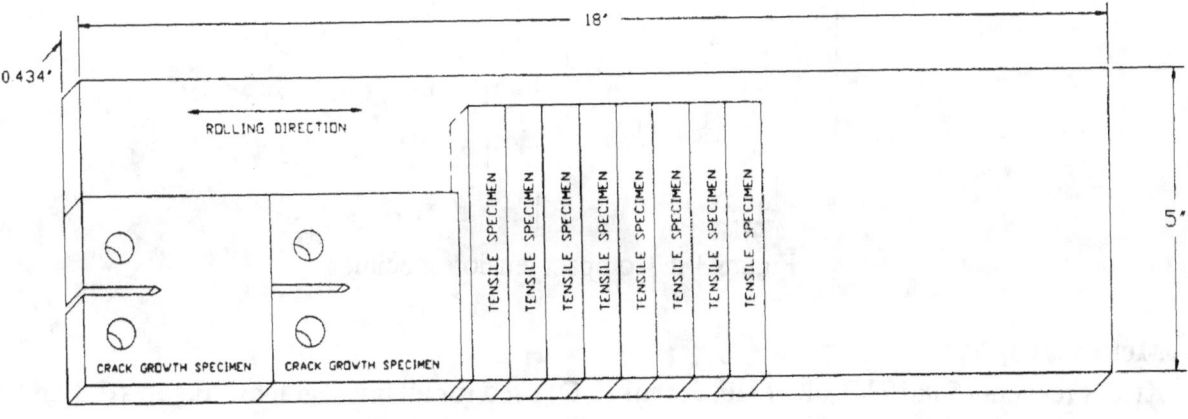

Figure 3-a. Pattern used to cut test specimens

Specimen Geometry:

Smooth tensile overload specimen dimensions are shown in Figure 3-b. Notched tensile overload specimen dimensions are shown in Figure 3-c. Compact tension specimens were made accor... ng to Figure 3-d.

51

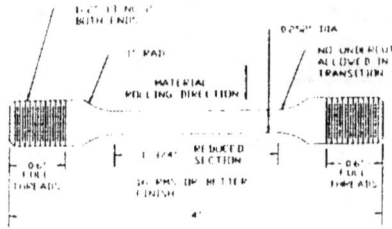 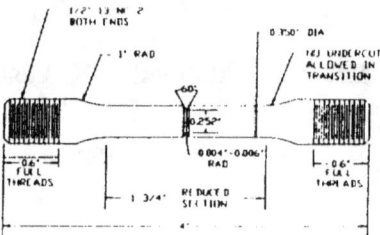

Figure 3-b. Smooth tensile overload specimen, 4" long with a 0.20" diameter.

Figure 3-c. Notched tensile overload specimen, 4 in long, 0.004-0.006" root radius, 0.20" notch diameter.

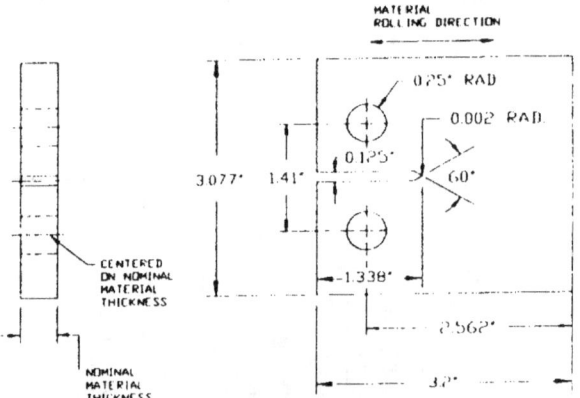

Figure 3-d. Compact tension specimen

Metallography:

Kroll's reagent (5ml HNO_3, 2ml HF, 100ml H_2O) used for all micrographs. Figures 3-1 and 3-2 show that the structure consists of slightly elongated grains of alpha (light) and intergranular beta (gray).

Heat Treatment:

As received annealed

TENSILE TEST DATA

Tensile Overload Smooth Sample at Room Temperature

Tensile strength 142,500 psi

Macroscopic Appearance of Fracture

The surface of this fracture is rough and has many distinct ridges as seen in Figure 3-3. The ridges are associated with the elongated grain structure and the hot working process.

Microscopic Appearance of Fracture

Figure 3-4 to 3-5 Area A: This region is on a shear lip. Although the surface is relatively smooth, ductile rupture is evident in the form of shallow equiaxed dimples.

Figures 3-6 to 3-7 Area B: Another shear lip. The ductile dimples are elongated and oriented in the same direction.

Figures 3-8 to 3-11 Area C: This area is representative of the central region of the sample. Note the ridges associated with the grain structure. There are numerous ductile dimples.

Tensile Overload Smooth Sample at 77 K
Tensile Strength 214,850 psi

Macroscopic Appearance

The macrofractograph of this fracture shown in Figure 3-12 displays a central region of rough surface with many small ridges oriented parallel to the rolling direction. The central region is surrounded by a circumferential shear region which exhibits a smoother texture.

Microscopic Appearance

Figures 3-13 to 3-16 Area A: This area is representative of the ductile overload found on most of the fracture surface. Notice that the fracture is more brittle than the room temperature sample. There is secondary cracking in this sample.

Figures 3-17 to 3-18 Area B: Fracture in this area is much like the fracture in the shear lip area of the room temperature sample. The dimples are shallower and smaller in size as compared to those in area A.

Tensile Overload Notched at Room Temperature
Tensile strength 210,700 psi

Macroscopic Appearance

Figure 3-19 shows the macrofractograph of this sample. The appearance is smooth with a faint radial pattern generating from the fracture origin.

<u>Microscopic Appearance</u>
Figures 3-20 to 3-23 Area A: These fractographs are typical of the morphology of the central part of the specimen. Dimple rupture is the predominant mode.
Figures 3-24 to 3-27 Area B: This area is near the fracture origin. The fracture is rougher and shows evidence of ductile tearing and secondary cracking.

Tensile Overload Notched at 77 K
Tensile strength 273,500 psi

<u>Macroscopic Appearance</u>
The macrofractograph of this fracture is shown in Figure 3-28. The surface is slightly rougher than the notched room temperature sample. The radial pattern originating from the fracture origin is also more pronounced.

<u>Microscopic Appearance</u>
Figures 3-29 to 3-31 Area A: These fractographs show the ductile rupture typical of the fracture. Note that there is less plasticity in this sample than in the room temperature notched sample as the dimples are less deep.

Figures 3-32 to 3-34 Area B: This area is near the fracture origin. Note that there are signs of lessened ductility such as secondary cracking.

HIGH CYCLE FATIGUE TEST DATA
Specimen was fatigued at room temperature in tensile - tensile loading

<u>Macroscopic Appearance</u>
Seven fatigue regions are shown in the optical photograph in Figure 3-35. Fatigue fracture Regions 2 through 6 were documented with fractographs. Region 1 and region 7 are the precrack and final fast fracture regions respectively, and they were not documented. The data depicting the test conditions used to grow the crack in Regions 2 through 6 is presented in the chart below, where R represents the ratio of the minimum to maximum load and K_{max} is the maximum stress concentration at the crack tip.

Region	R	K_{max} (ksi-in$^{1/2}$)	da/dN
2	0.75	20	9.6×10^{-7}
3	0.50	20	6.9×10^{-6}
4	0.25	20	1.9×10^{-5}
5	0.10	20	2.0×10^{-5}
6	0.75	40	not available

Microscopic Appearance

Figures 3-36 to 3-39 Region 2: The striations are difficult to find even at 10,000X with these load parameters.

Figures 3-40 to 3-44 Region 3: Striations are more evident as a smaller load ratio is applied.

Figures 3-45 to 3-48 Region 4: The striations are more pronounced as we apply an even smaller load ratio.

Figures 3-49 to 3-52 Region 5: The fractographs resemble those for area 4.

Figures 3-53 to 3-56 Region 6: In this region the K_{max} value is increased but the R value is the same as Regions 2 thru 5. Note that the striation spacings are about the same size as those in Region 3.

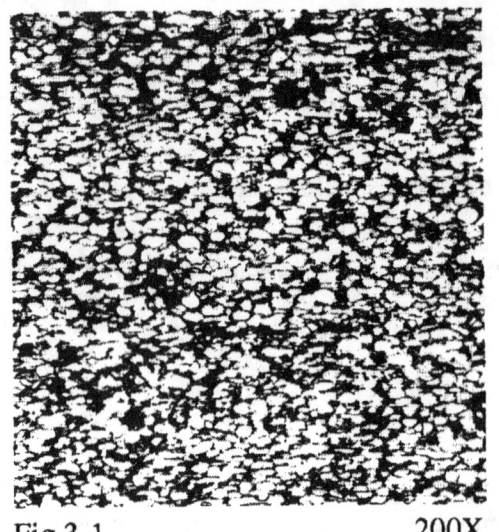

Fig 3-1   200X

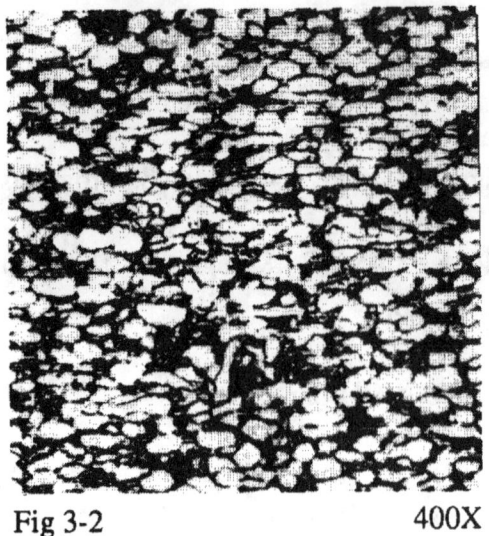

Fig 3-2   400X

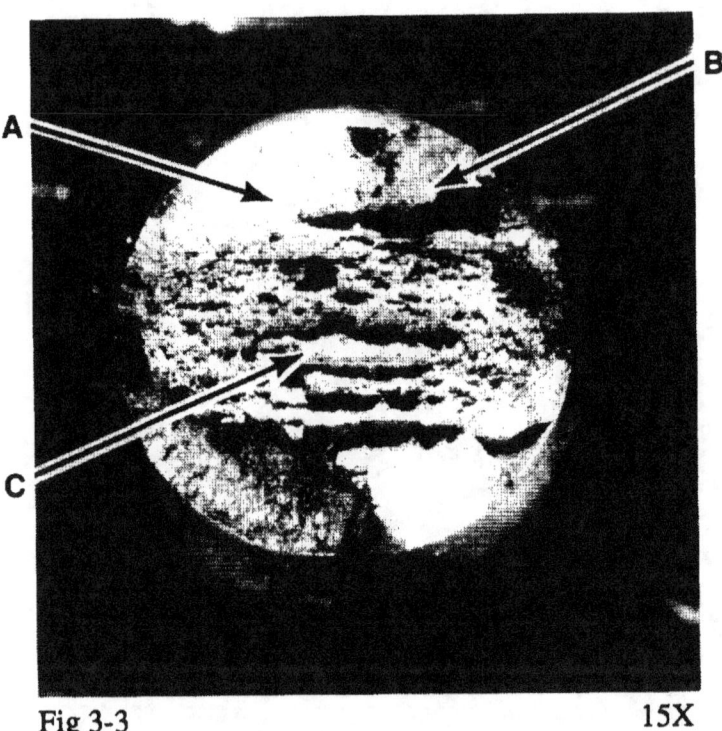

Fig 3-3   15X
Optical photomacrograph of fracture surface

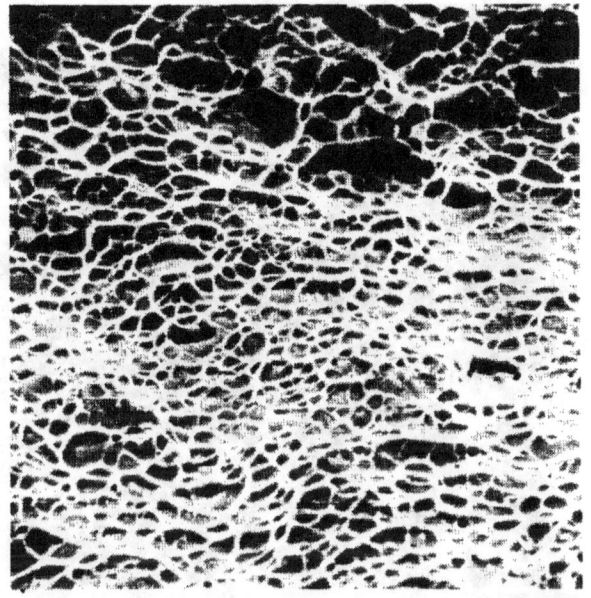

Fig 3-4   640X.
Region A. Shear lip area.

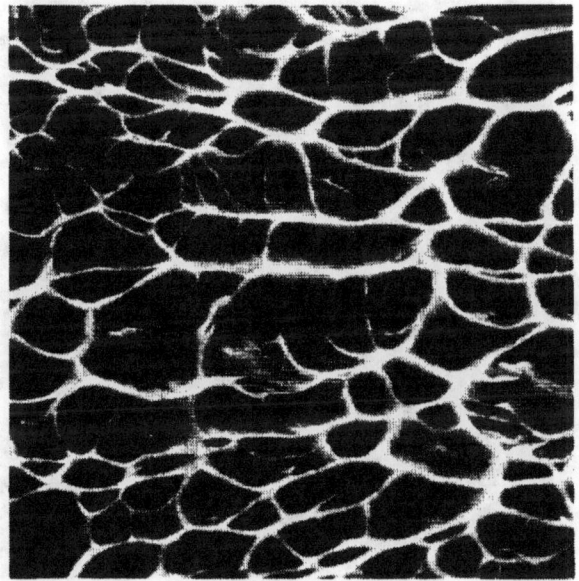

Fig 3-5   2500X
Region A. Enlarged view of center of Fig. 3-4.

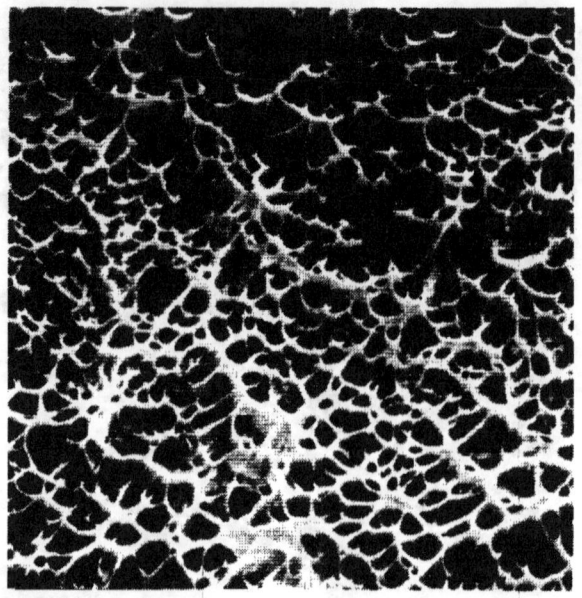

Fig 3-6   640X
Region B. Another shear area.

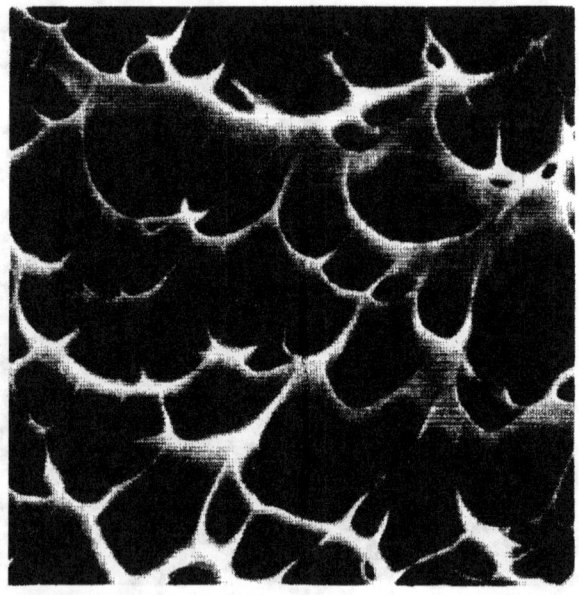

Fig 3-7   2500X
Region B. Enlarged view of center of Fig. 3-6.

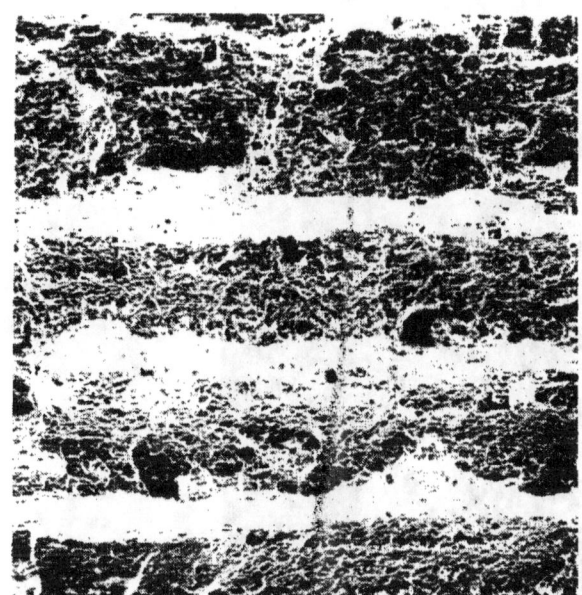

Fig 3-8   80X
Region C. Central region of fracture.

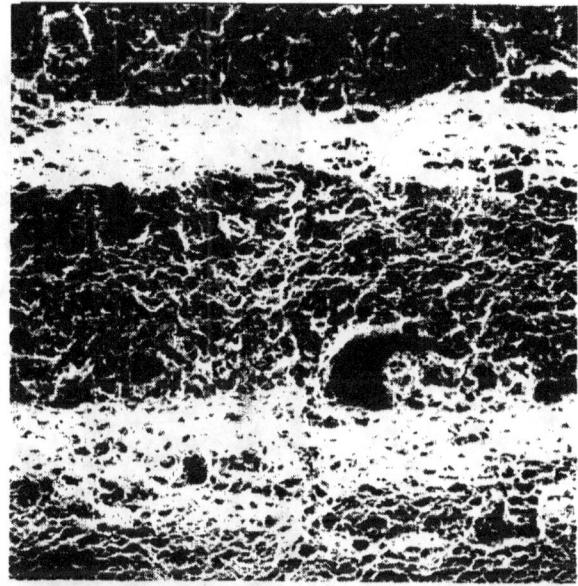

Fig 3-9   160X
RegionC. Enlarged view of the central region
of fracture.

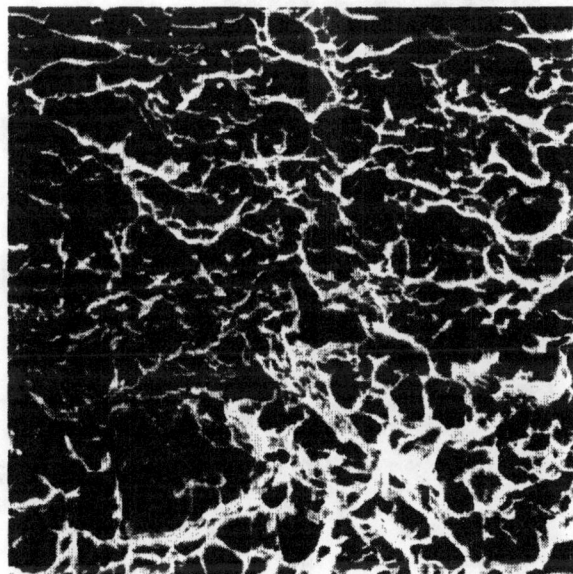

Fig 3-10   640X
Region C. Enlarged view of the center
of Fig. 3-9.

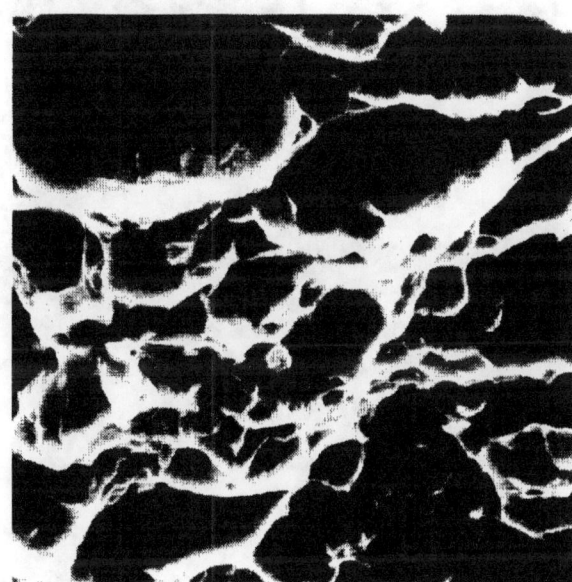

Fig 3-11   2500X
Region C. Enlarged view of the center
of Fig. 3-10.

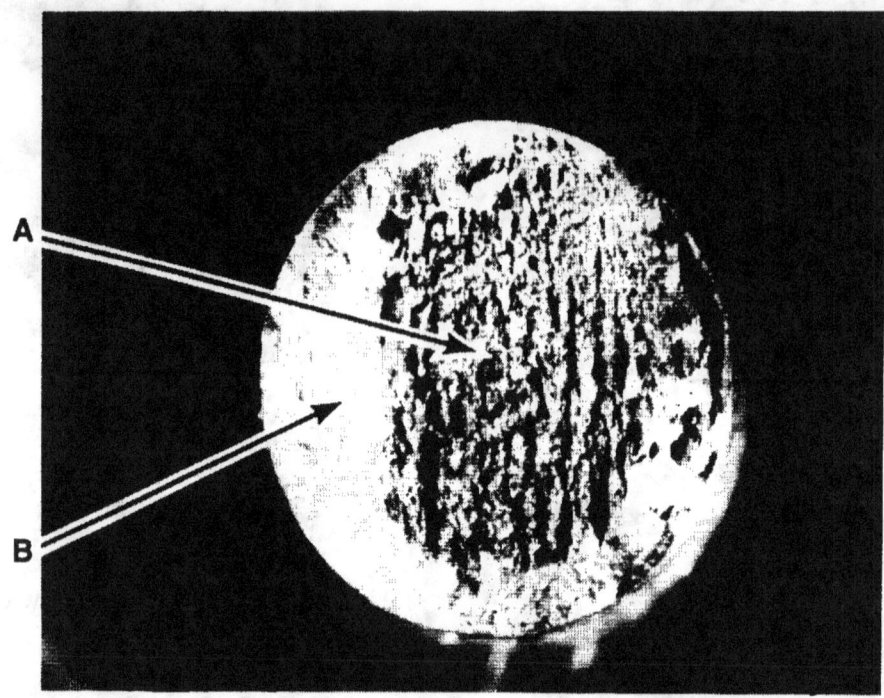

Fig 3-12   15X
Optical photomacrograph of fracture surface.

60

Fig 3-13   160X
Region A. Central region of fracture.

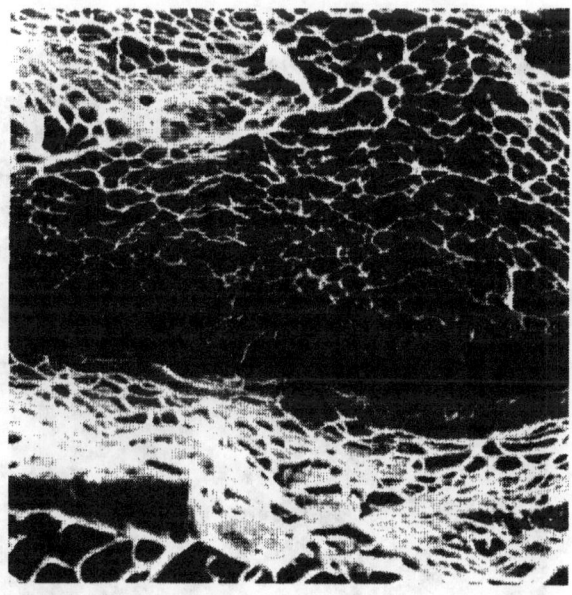

Fig 3-14   640X
Region A. Enlarged view of Fig. 3-13.

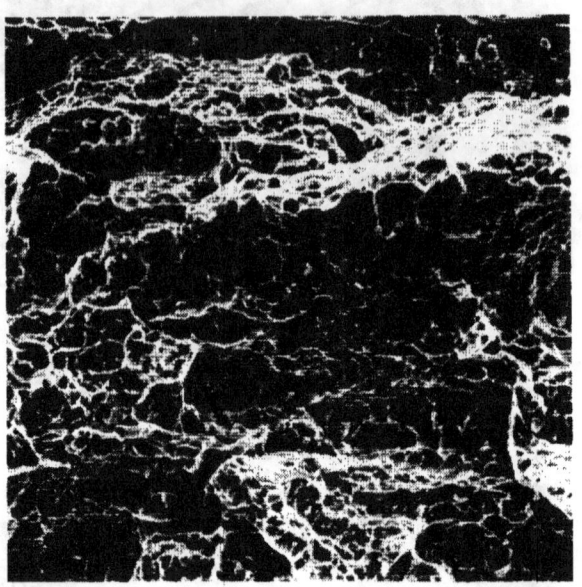

Fig 3-15   320X
Region A. Another area in the central region
of fracture.

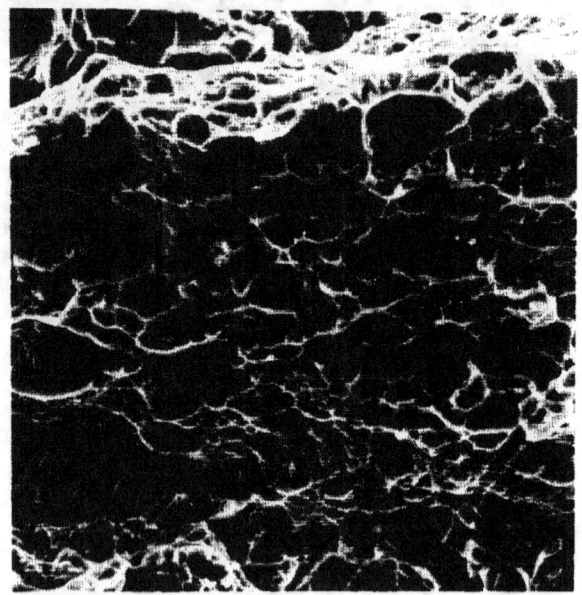

Fig 3-16   640X
Region A. Enlarged view of Fig. 3-15.

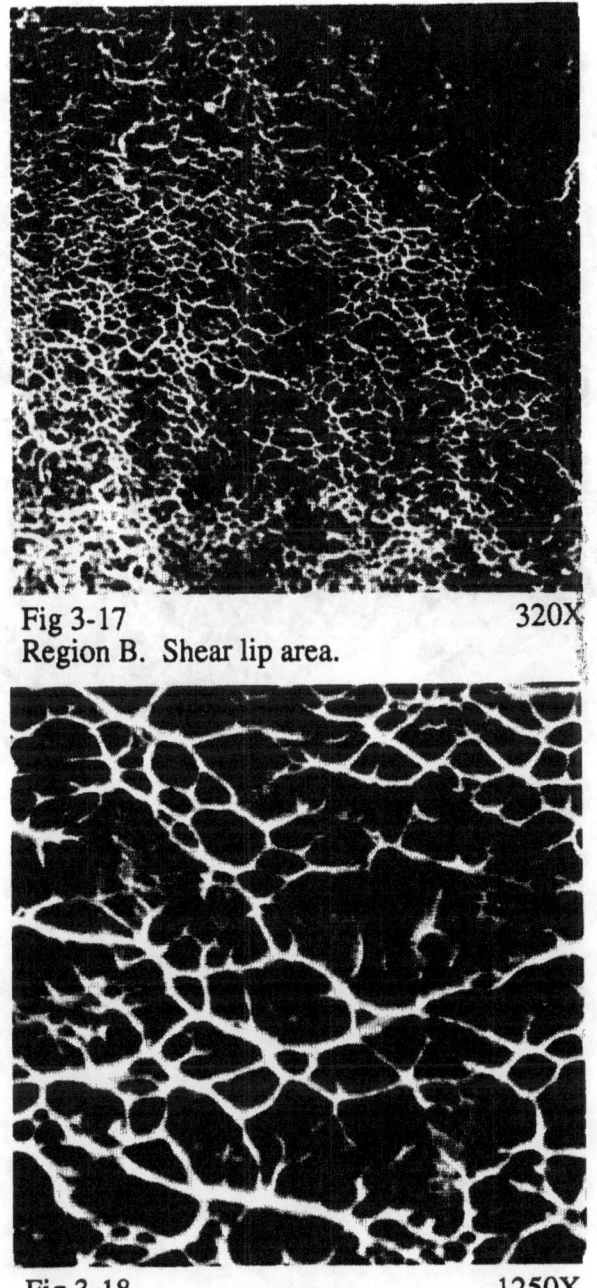

Fig 3-17   320X
Region B. Shear lip area.

Fig 3-18   1250X
Region B. Enlarged view of Fig. 3-17.

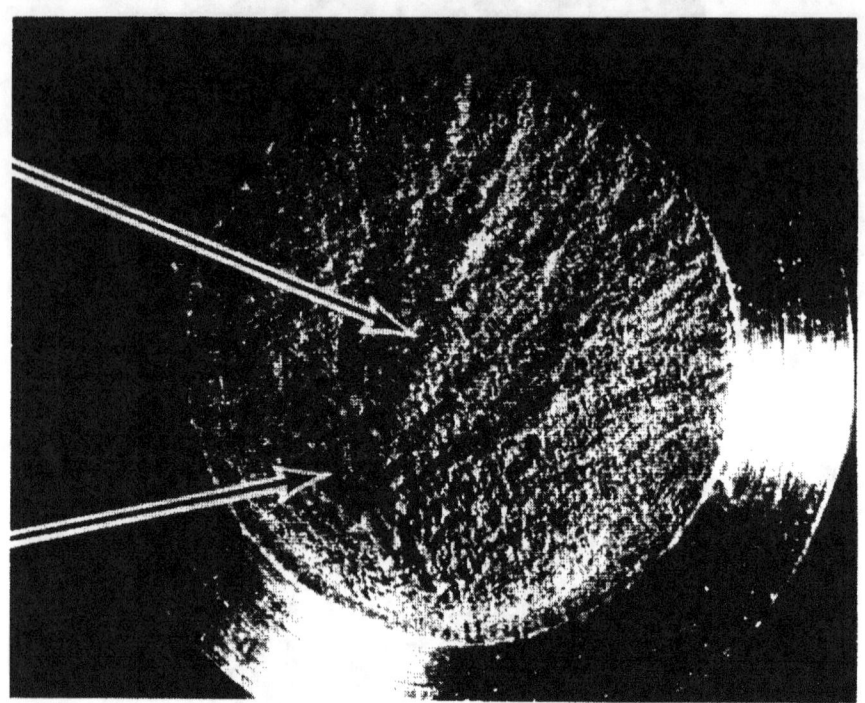

Fig 3-19   15X
Optical photomacrograph of fracture surface.

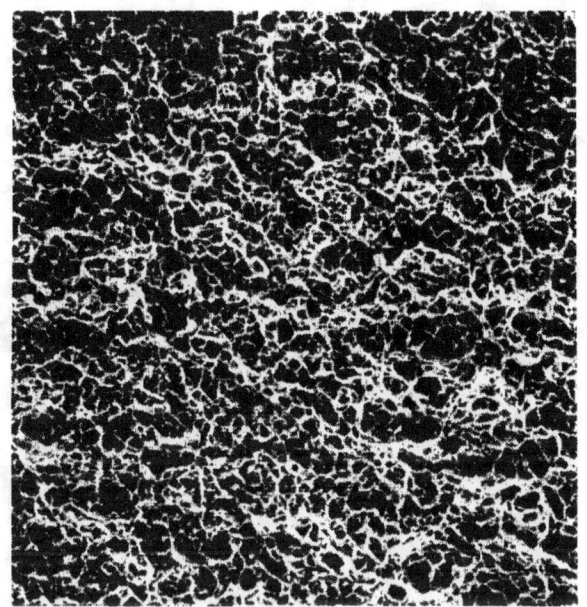

Fig 3-20   160X
Region A. Central region of fracture.

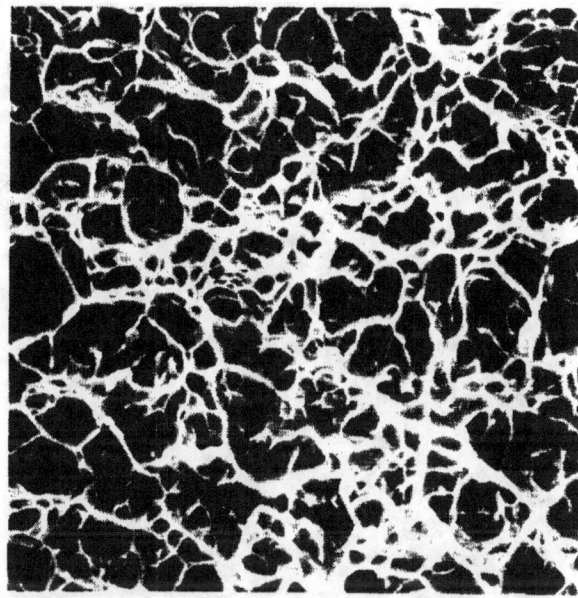

Fig 3-21   640X
Region A. Enlarged view of Fig. 3-20.

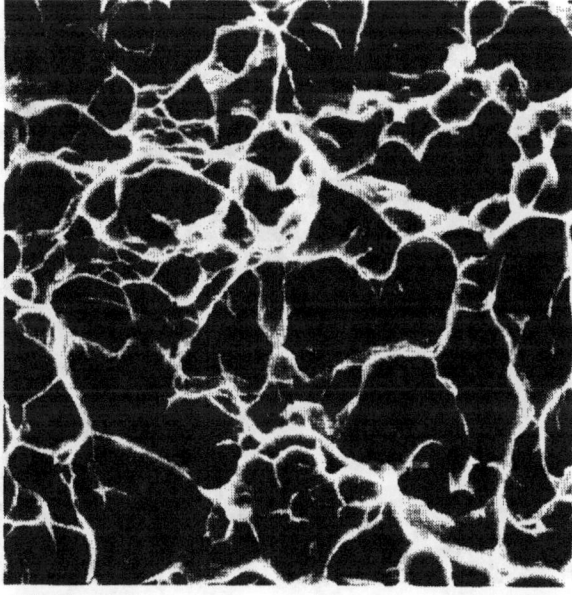

Fig 3-22   1250X
Region A. Enlarged view of Fig. 3-21.

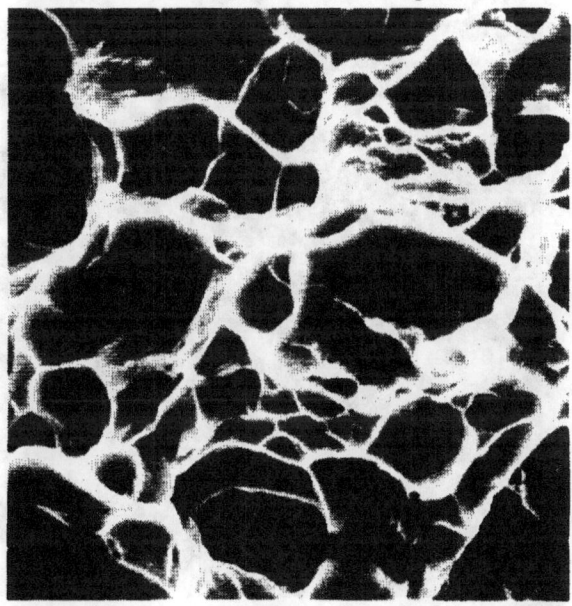

Fig 3-23   2500X
Region A. Enlarged view of Fig. 3-22.

Fig 3-24   160X
Region B. Area near the fracture origin.

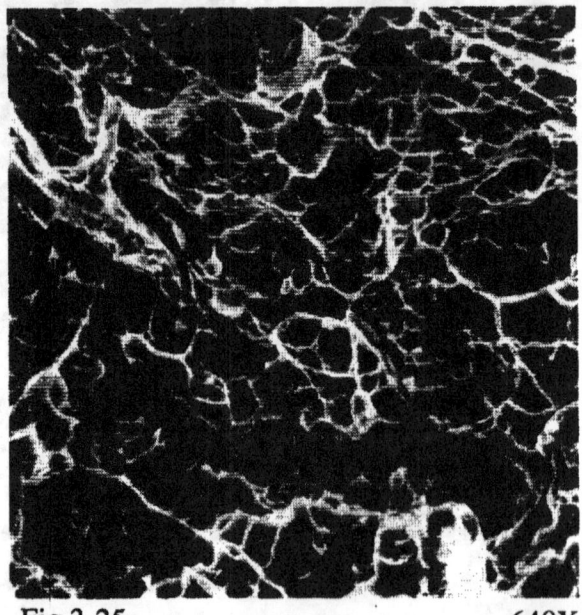

Fig 3-25   640X
Region B. Enlarged view of Fig. 3-24.

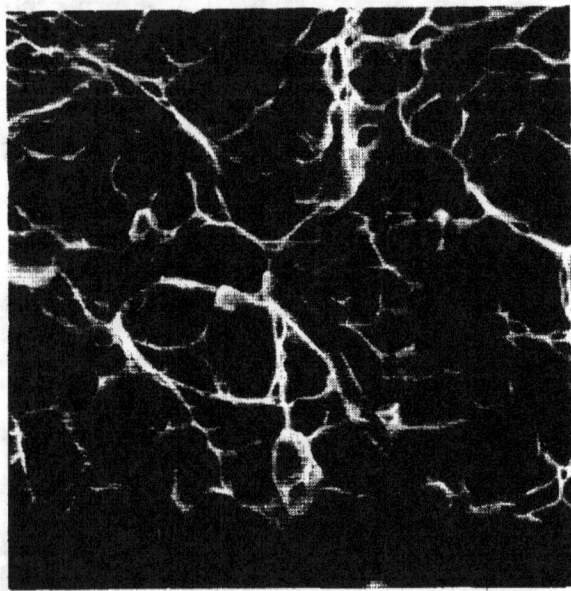

Fig 3-26   1250X
Region B. Enlarged view of Fig. 3-25.

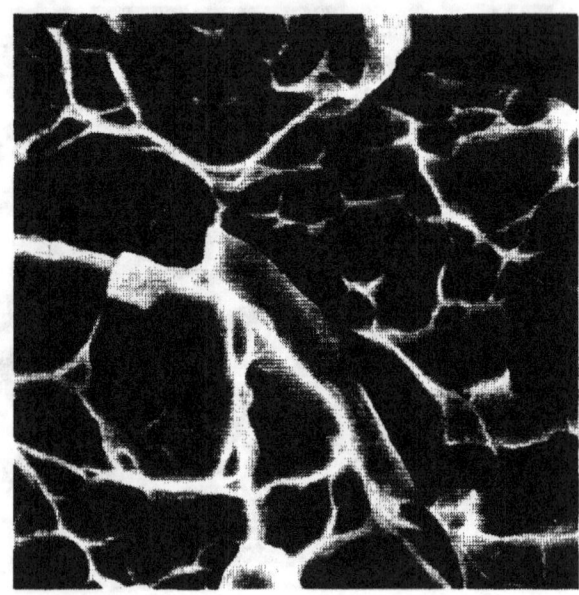

Fig 3-27   2500X
Region B. Enlarged view of Fig. 3-26.

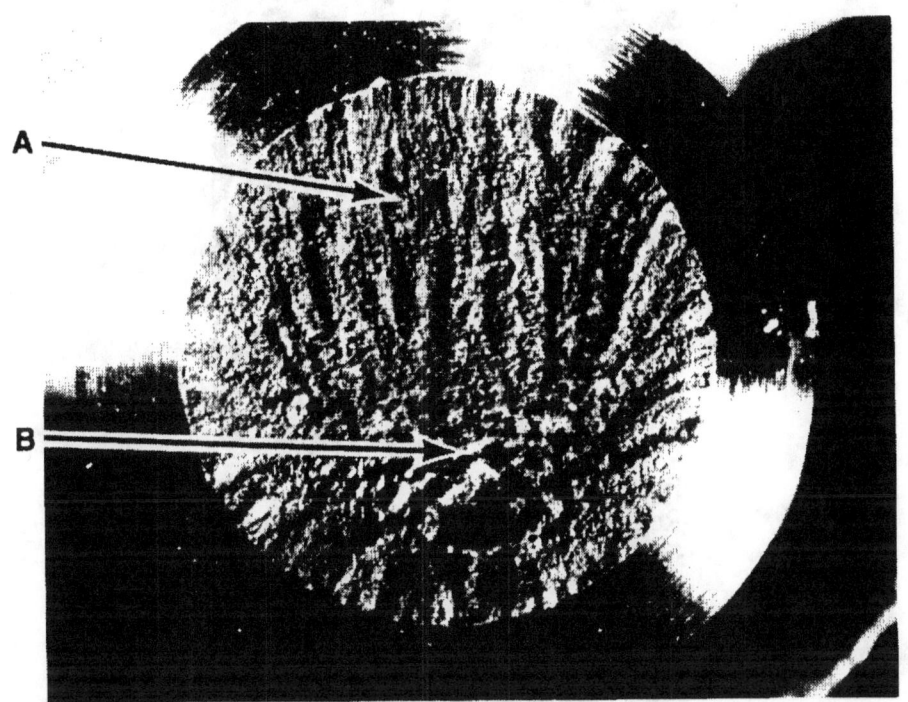

Fig 3-28
Photomacrograph of fracture surface.

15X.

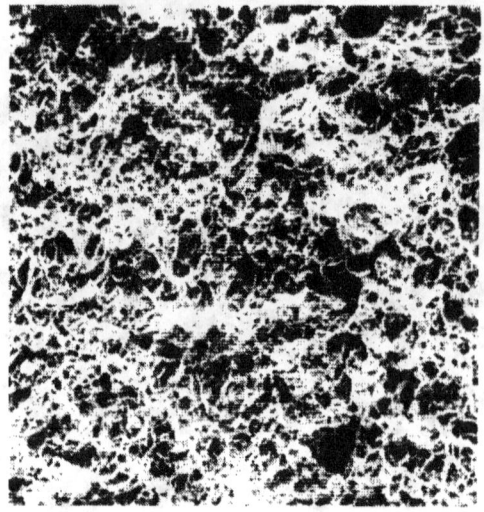

Fig 3-29   160X
Region A. Central region of fracture.

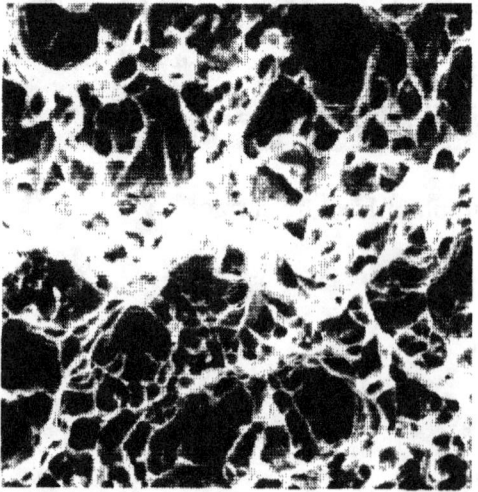

Fig 3-30   640X
Region A. Enlarged view of Fig. 3-29.

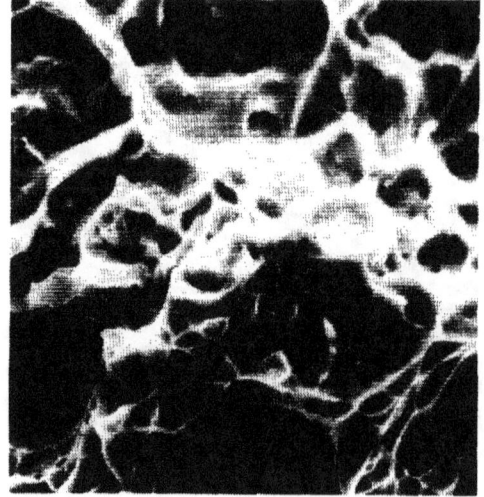

Fig 3-31   1250X
Region A. Enlarged view of Fig. 3-30.

Fig 3-32   160X
Region B. Region near the fracture origin.

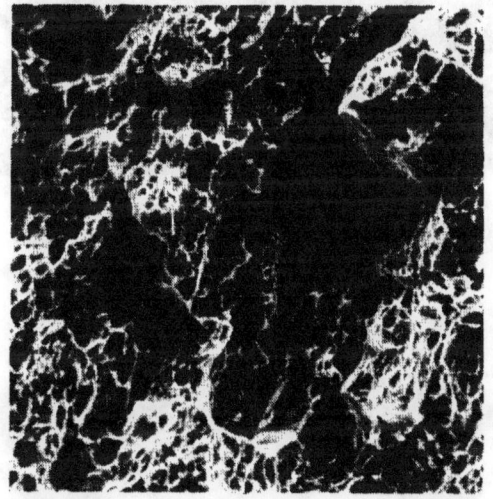

Fig 3-33   320X
Region B. Enlarged view of Fig. 3-32.

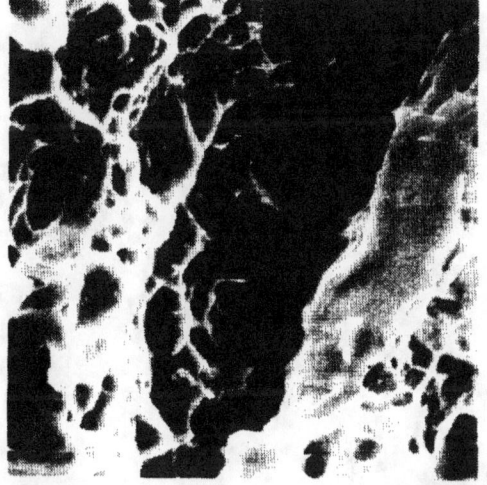

Fig 3-34   1250X
Region B. Enlarged view of Fig. 3-33.

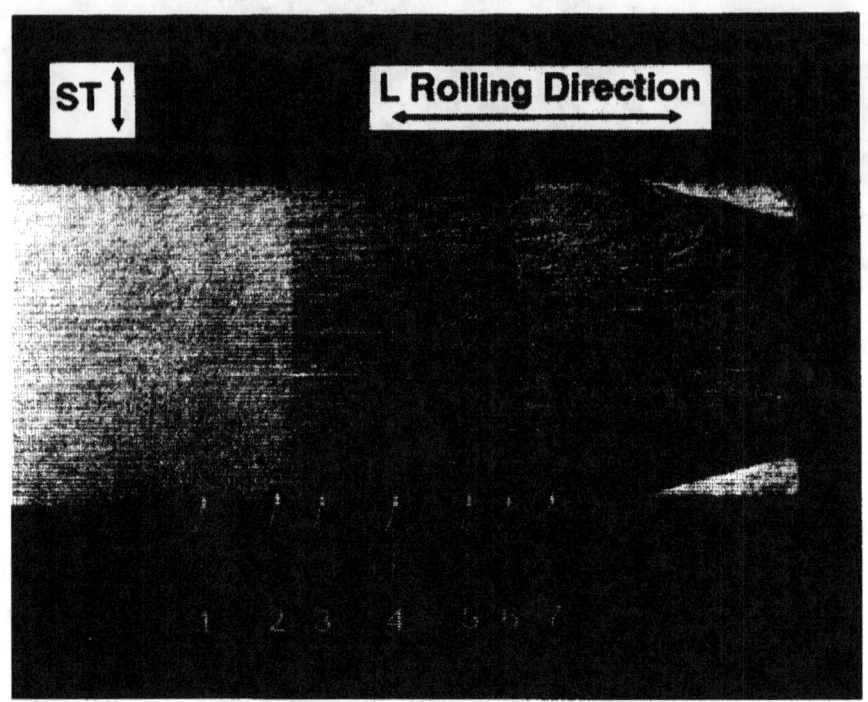

Fig. 3-35   4X
Optical photomacrograph of fatigue fracture surface.

Fig. 3-36   640X
Region 2. Faint fatigue striations.

Fig. 3-37   1250X
Region 2.

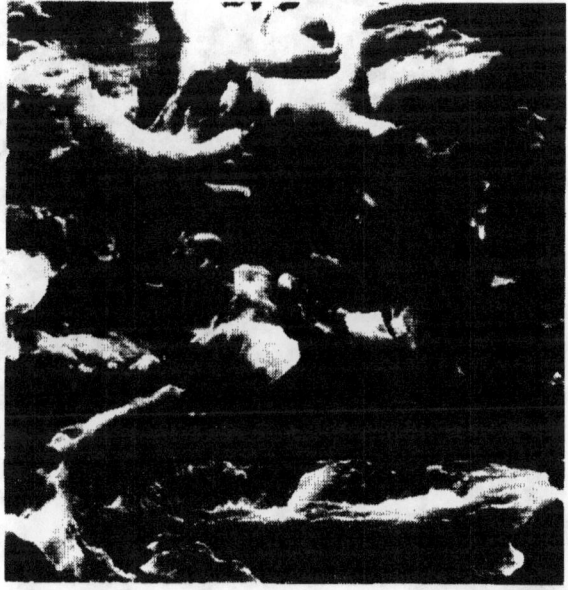

Fig. 3-38   2500X
Region 2.

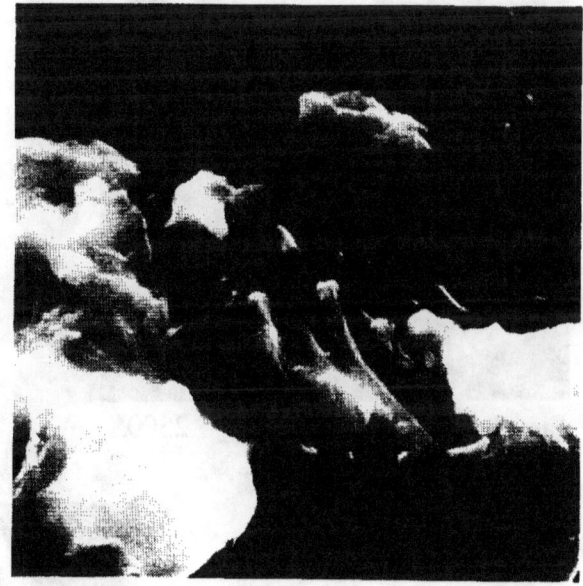

Fig. 3-39   10000X
Region 2.

71

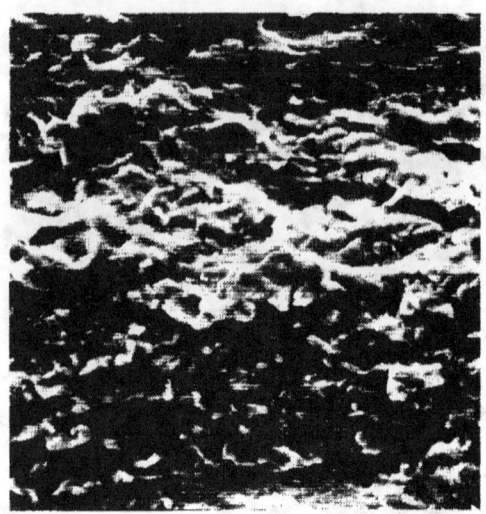

Fig. 3-40   640X
Region 3. Fatigue striations.

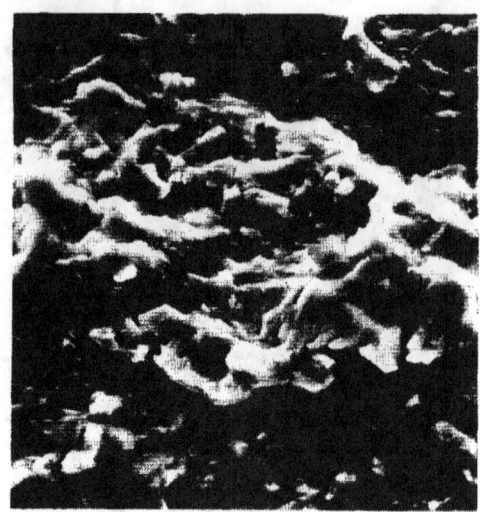

Fig. 3-41   1250X
Region 3.

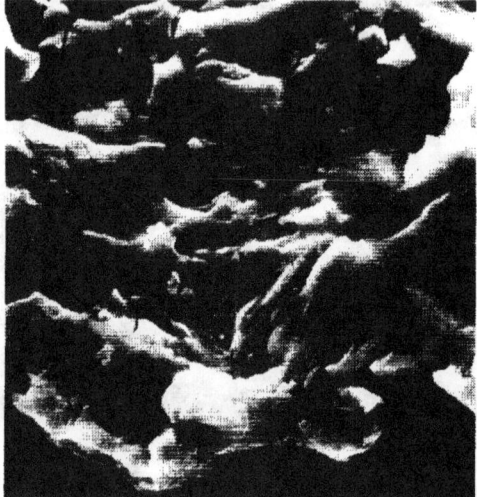

Fig. 3-42   2500X
Region 3.

Fig. 3-43   5000X
Region 3

Fig. 3-44   10000X
Region 3

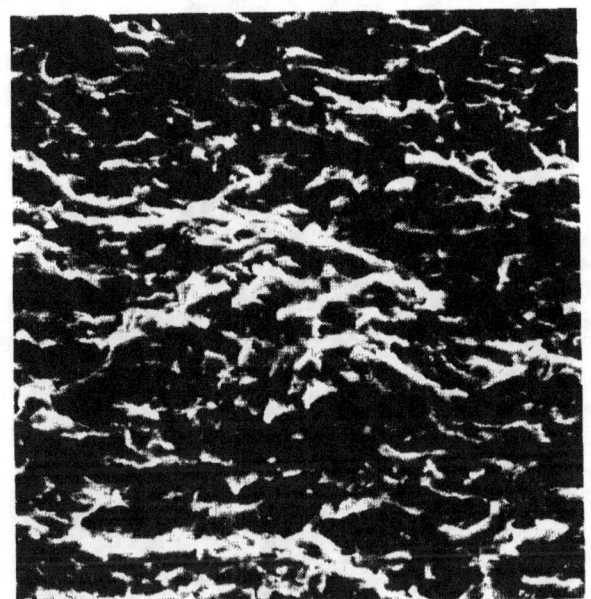

Fig. 3-45   640X
Region 4. Fatigue striations.

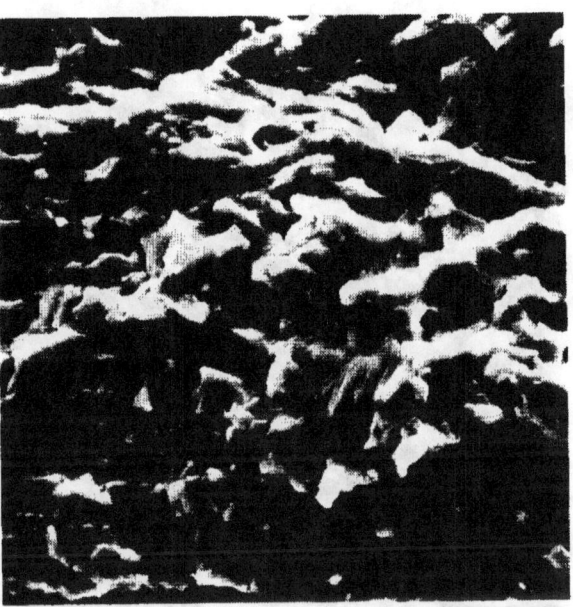

Fig. 3-46   1250X
Region 4.

Fig. 3-47   2500X
Region 4.

Fig. 3-48   5000X
Region 4.

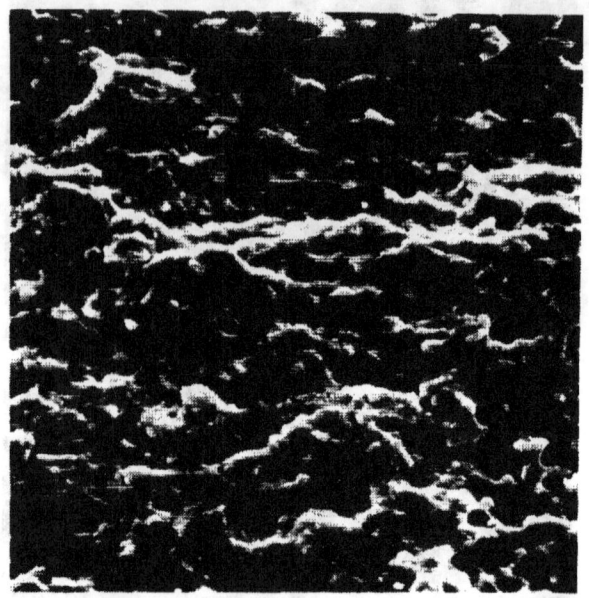

Fig. 3-49   640X
Region 5. Fatigue striations.

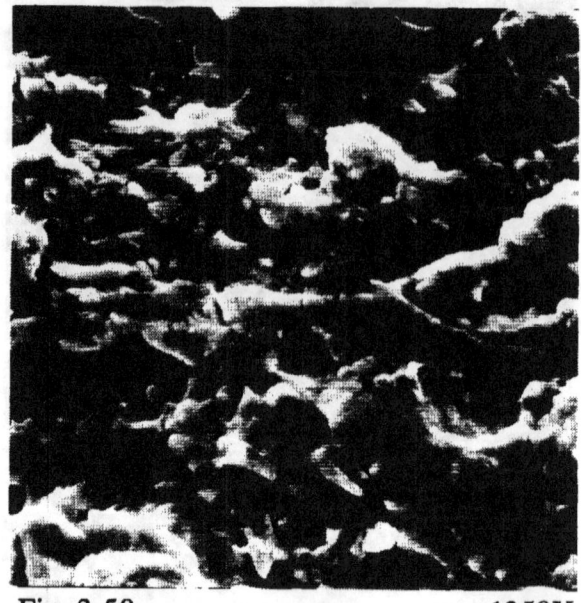

Fig. 3-50   1250X
Region 5.

Fig. 3-51   2500X
Region 5.

Fig. 3-52   5000X
Region 5.

Fig. 3-53   640X
Region 6. Fatigue striations.

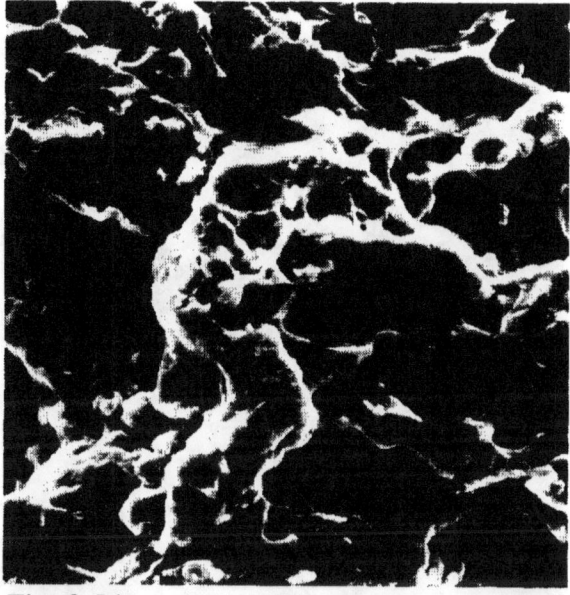

Fig. 3-54   1250X
Region 6.

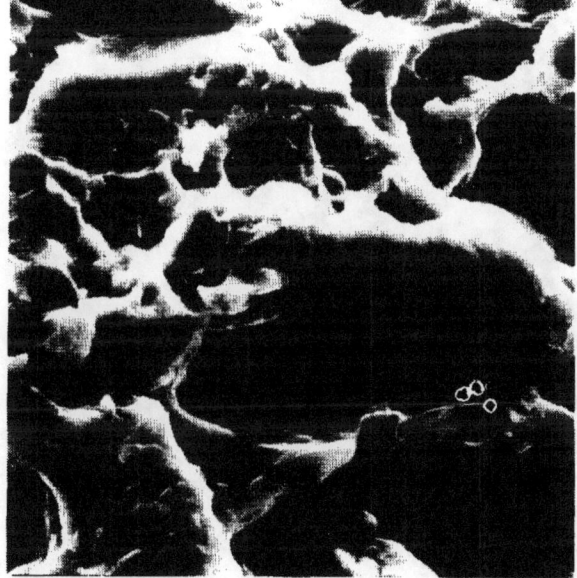

Fig. 3-55   2500X
Region 6.

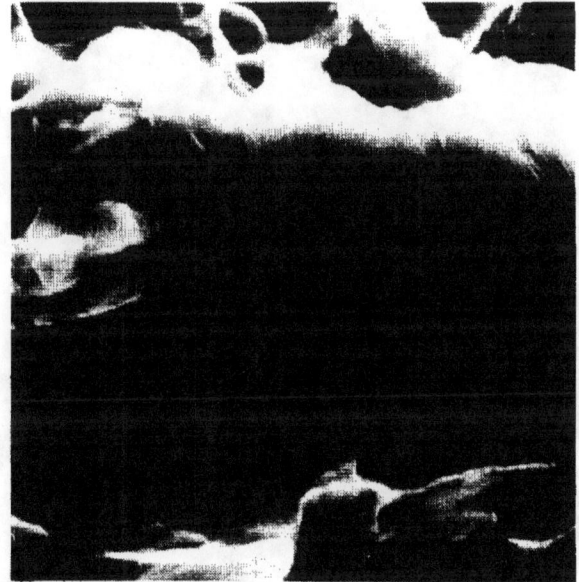

Fig. 3-56   5000X
Region 6.

Specimen 4   Copper Beryllium CA172 Wrought

Composition:

Element	Nominal weight percent	Actual weight percent
Be	1.8-2.0	1.96
Co	0.20 min	0.238
Cu	Balance	Balance

Supplier: Cabot Wrought Product (now known as NGK Metals Corporation, Reading, PA, 19612-3367)

Specimen Condition:

Heat treatment: TF00 - Precipitation hardened at 600 F for 3 hours Hardness: HRC 38.
Dimensions: 1.25" Tk X 5.00" W X 18.0" L plate. Test specimens were cut from the plate according to the diagram in Figure 4-a.

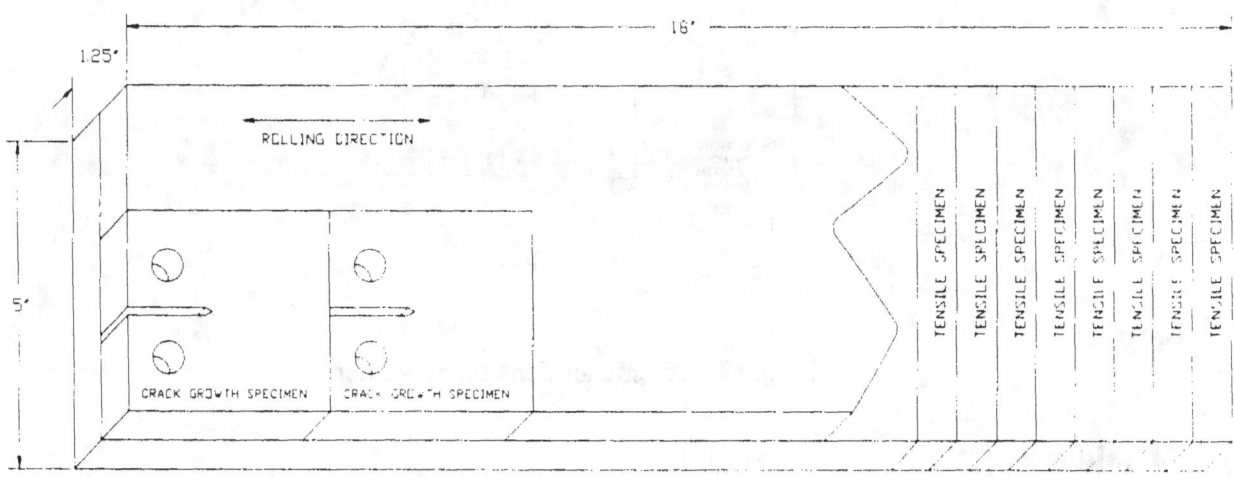

Figure 4-a. Pattern used to cut test specimens.

Specimen Geometry:

The dimensions for the smooth tensile overload specimens are shown in Figure 4-b, and the dimensions for the notched tensile overload specimens are shown in Figure 4-c. The compact tension specimen dimensions are shown in Figure 4-d.

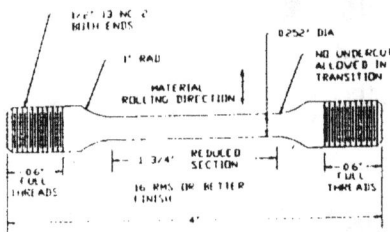

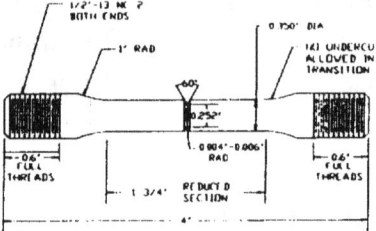

Figure 4-b. Smooth tensile overload specimen, 4" long, 0.252" diameter.

Figure 4-c. Notched tensile overload specimen, 4" long, 0.004-0.006" root radius, 0.252" notch diameter.

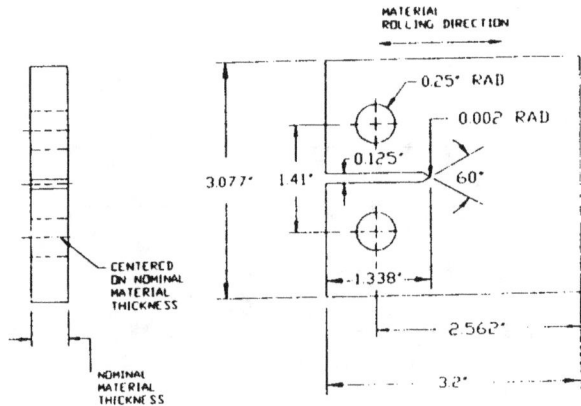

Figure 4-d. Compact Tension Specimen

Metallography:

Specimen was etched with Copper Etch (4 ml Saturated NaCl, 8 ml H_2SO_4, 2 gms Potassium Dichromate, 100 ml H_2O). The microstructure contains a dark colored gamma phase in the grain boundaries, with uniformly dispersed dark cobalt-beryllide and light primary beta particles also present. Note the twin formation due to working of the metal. The low magnification photograph in Figure 4-3 shows large bands distributed in the microstructure. These bands are also visible in the fracture surfaces.

TENSILE TEST DATA

Tensile Overload Smooth At Room Temperature
Tensile Strength 174,300 psi Yield Strength 113,000 psi

Macroscopic Appearance of Fracture
The fracture is shown is Figure 4-4. The overall appearance of this fracture is smooth in the center with radial lines originating from the fracture origin. Several shear lips are located circumferentially.

Microscopic Appearance
Figures 4-5 to 4-11 Area A: These fractographs are representative of the fracture morphology in the center of the sample. There is transgranular fracture, secondary cracking, and cracking along the long bands that are seen in the microstructure (Fig 4-5 to 4-7). Dimple rupture is evident at both low and high magnification. Dimples are seen on the secondary crack walls and on facet surfaces. The facets are indicated with arrowheads in Figures 4-9, 4-10 and 4-11.

Figures 4-12 to 4-15 Area B: Long cracks and separations on the fracture surface occur at long bands that are seen in the microstructure.

Tensile Overload Smooth at 77 K
Tensile Strength 196,200 psi

Macroscopic Appearance:
This fracture is shown in Figure 4-16. The surface appears coarser than that of the room temperature sample. The radial lines are also more defined.

Microscopic Appearance
Figures 4-17 to 4-21 Area A: Note that this fracture surface looks similar to the room temperature fracture but exhibits more intergranular cracking.

Figures 4-22 to 4-24 Area B: These are fractographs in another area near the center of the specimen. Note the mixed mode fracture which is mostly transgranular but includes some intergranular cracking (Fig 4-22, 4-23) and the presence of ductile dimples at high magnification (Fig 4-24).

Tensile Overload of Notched Sample at Room Temperature
Tensile Strength 113,000 psi

<u>Macroscopic Appearance</u>

The fracture is shown in Figure 4-25. This specimen exhibits a smooth surface with small ridges in a slightly radial pattern.

<u>Microscopic Appearance</u>

Figures 4-26 to 4-29 Area A: This area is near the fracture origin. There is some intergranular fracture located among the transgranular ductile dimples.

Figures 4-30 to 4-31 Area B: Center of the sample. Mixed mode fracture including intergranular and transgranular fracture is found in this area also.

Tensile Overload of Notched Sample at 77 K
Tensile Strength 169,800 psi

<u>Macroscopic Appearance</u>

The fracture surface is in Figure 4-32. The cold temperature specimen exhibits a more pronounced radial pattern originating from the fracture origin and a slightly coarser fracture surface.

<u>Microscopic Appearance</u>

Figure 4-33 to 4-35 Area A: This area is near the fracture origin. The surface contains intergranular and secondary cracking. The surface reflects the grain structure as more intergranular and secondary cracking has taken place. Note that small ductile dimples are still present throughout the sample (Fig 4-35).

Figure 4-36 to 4-38 Area B: A location near the center of the sample that contains a large secondary crack and several smaller secondary cracks (Fig 4-37, 4-38).

HIGH CYCLE FATIGUE TEST DATA

<u>Macroscopic Appearance</u>

The different regions of crack growth rates are not visible in the optical photograph in Figure 4-39 but a scribe mark is identified as the approximate beginning of each region. Region 1 and Region 6 are the precrack and final fast fracture regions, respectively, and were not documented. Regions 2 through 5 represent four different loading conditions and their fracture surfaces were documented with SEM fractographs.

The chart below lists the test conditions for the documented regions.

Region	R	K_{max} (ksi-in$^{1/2}$)	dA/dN
2	0.1	10	1.6×10^{-6}
3	0.1	12.5	2.2×10^{-6}
4	0.1	13.75	2.6×10^{-6}
5	0.1	15	3.5×10^{-6}

Microscopic Appearance

Figures 4-40 to 4-43 Region 2: The striations are very difficult to find even at 10,000X at this load parameter.

Figures 4-44 to 4-46 Region 3: The striations are not apparent at these load conditions either. The parallel markings are possibly slip planes.

Figures 4-47 to 4-49 Region 4: Striations are not apparent here either.

Figures 4-50 to 4-53 Region 5: Striations are evident at 5,000X and 10,000X.

Figures 4-54 to 4-56 Region 5: Another area in Region 5. Striations are not evident in this area.

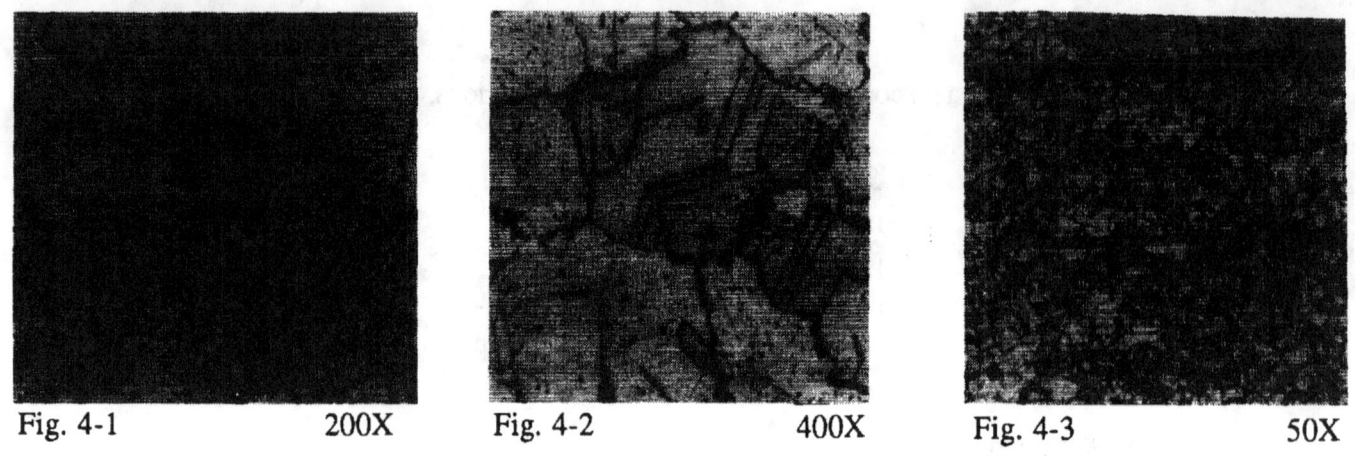

Fig. 4-1   200X   Fig. 4-2   400X   Fig. 4-3   50X

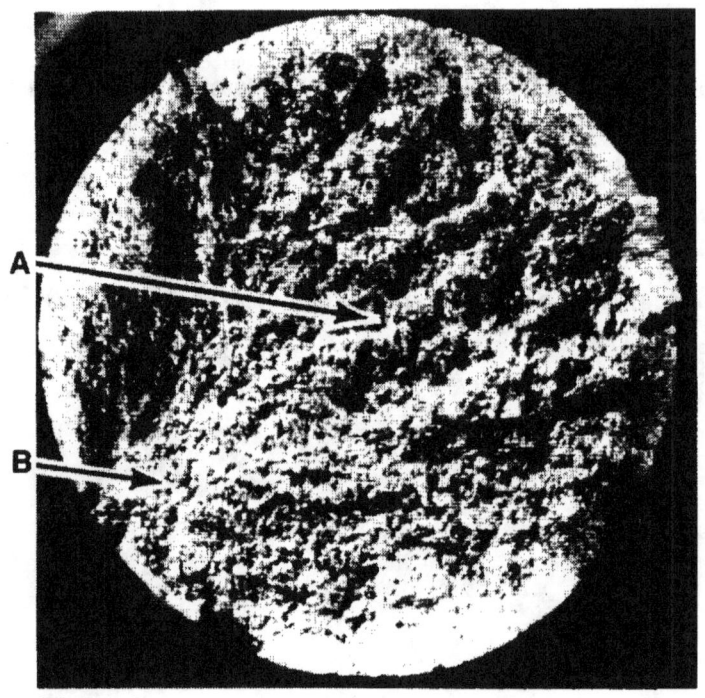

Fig. 4-4   14X
Optical photomacrograph of fracture surface.

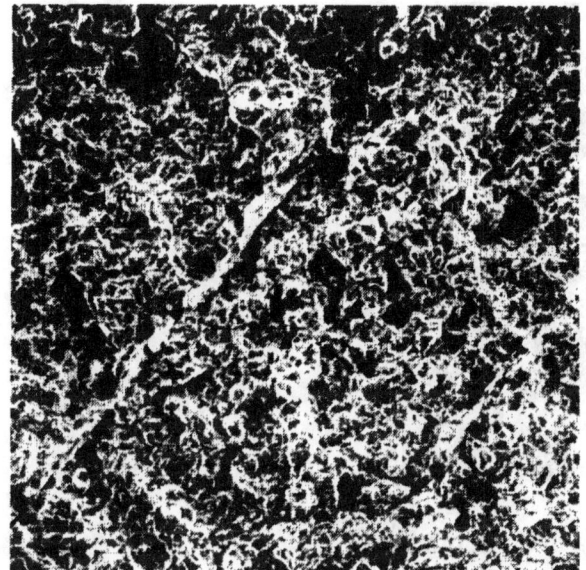

Fig. 4-5   40X
Region A. Central area of fracture.

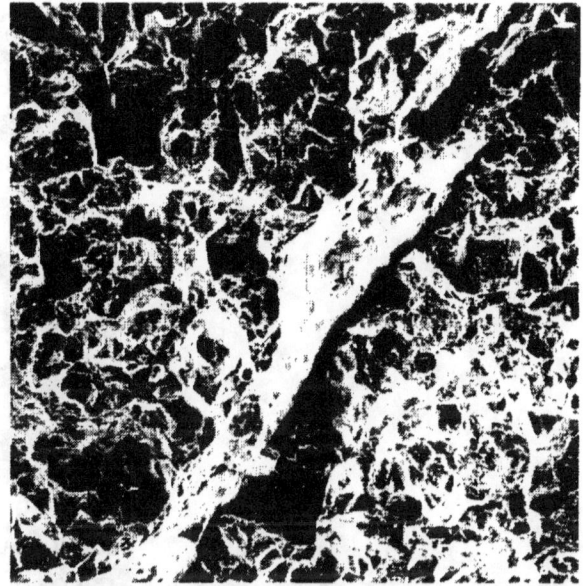

Fig. 4-6   160X
Region A. Secondary cracking.

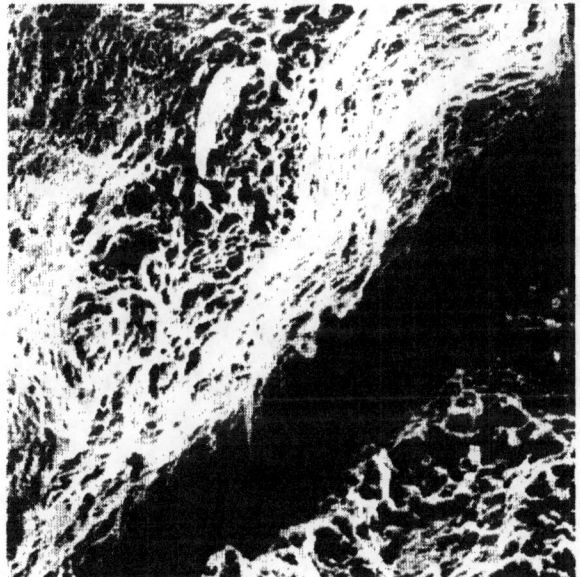

Fig. 4-7   1250X
Region A. Enlarged view of Fig. 4-6.

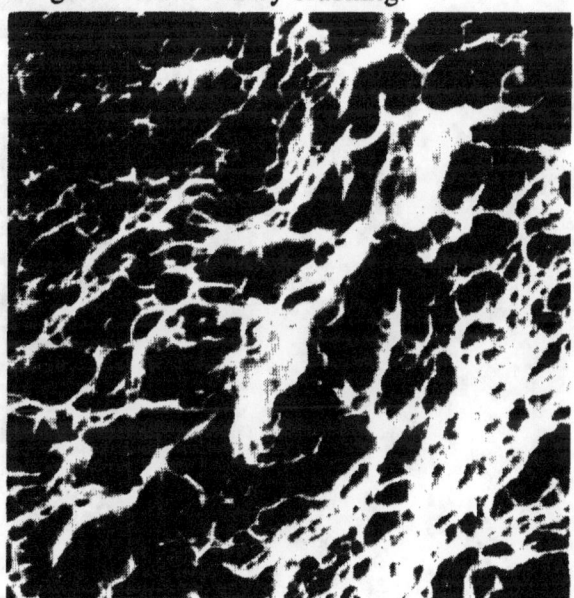

Fig. 4-8   5000X
Region A. Microdimples on the crack surface.

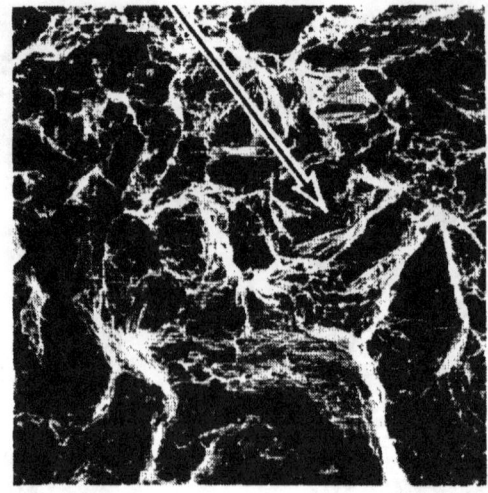

Fig. 4-9   640X
Region A. Another area in the central
region of fracture.

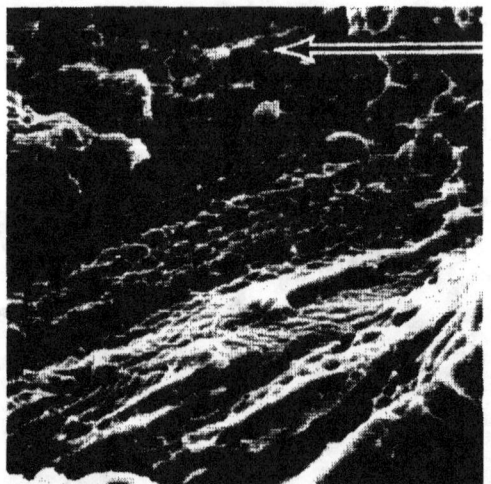

Fig. 4-10   5000X
Region A. Microdimples on a grain facet.

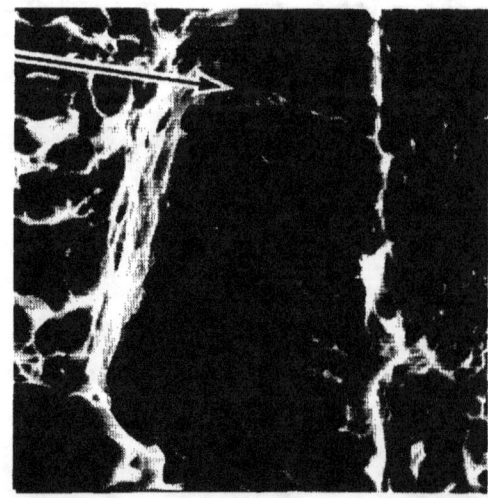

Fig. 4-11   5000X
Region A. Microdimples on a grain facet.

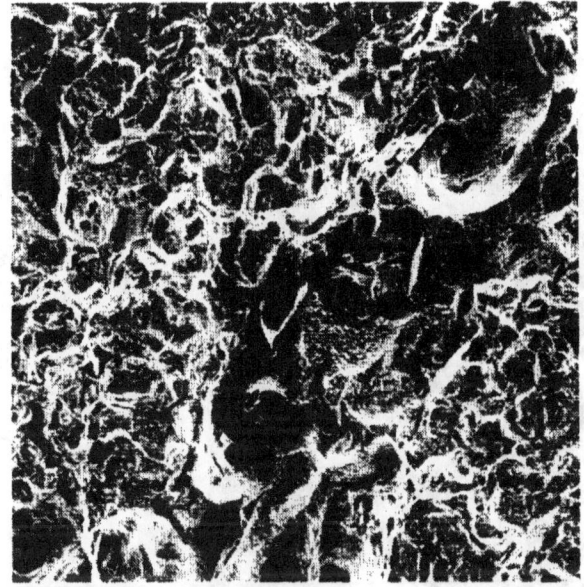

Fig. 4-12   160X
Region B. Secondary crack.

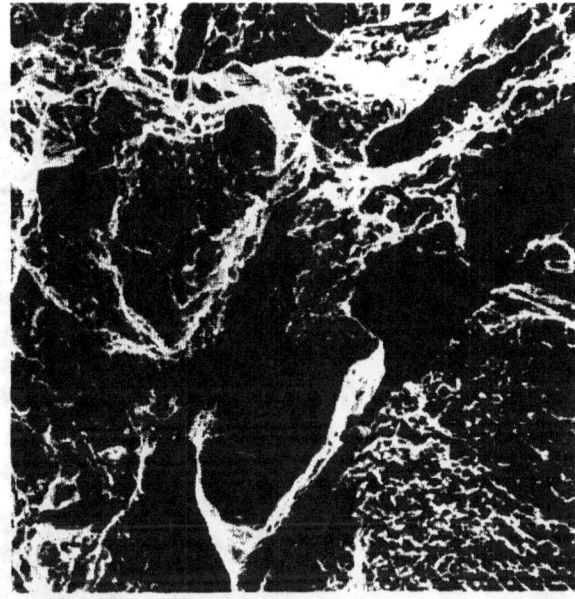

Fig. 4-13   640X
Region B. Enlarged view of Fig. 4-12.

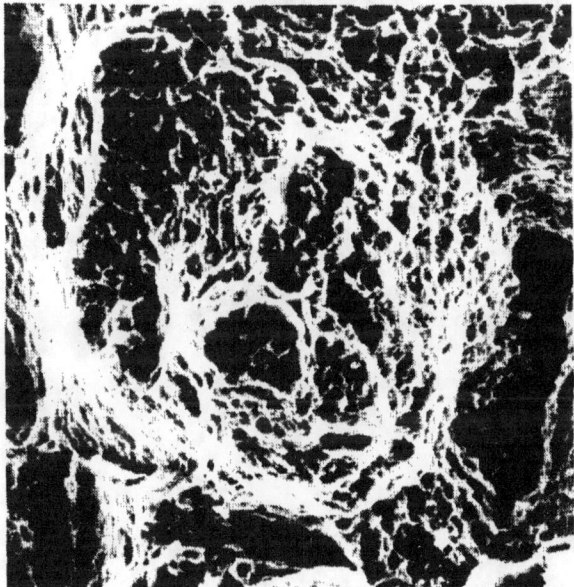

Fig. 4-14   1250X
Region B. Enlarged view of 4-13.

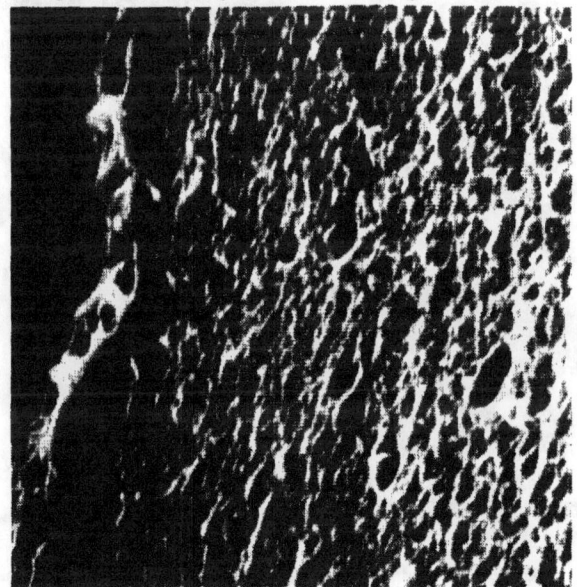

Fig. 4-15   5000X
Region B. Microdimples on right center

grain surface in Fig. 4-14.

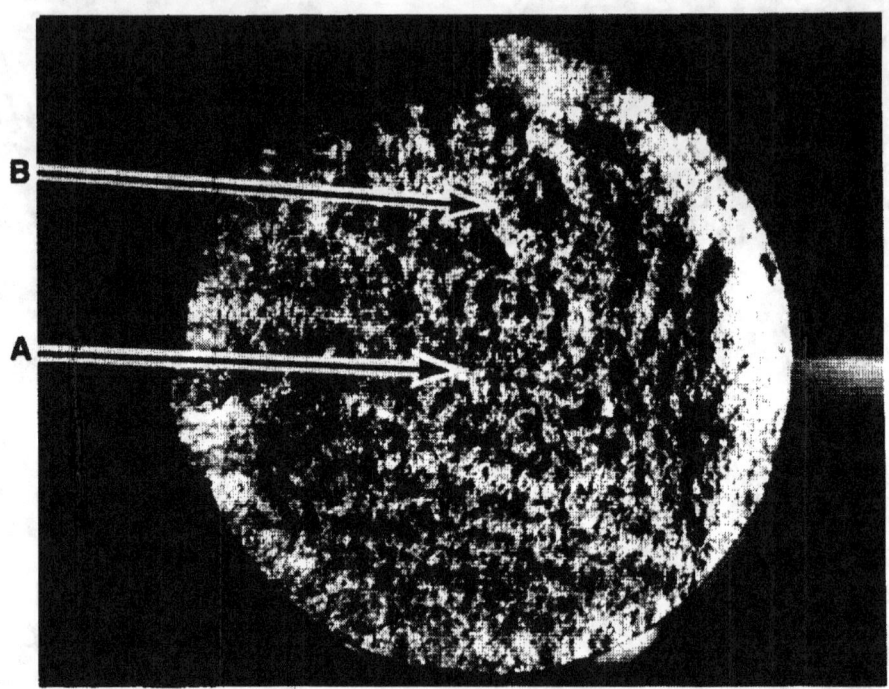

Fig. 4-16   14X

86

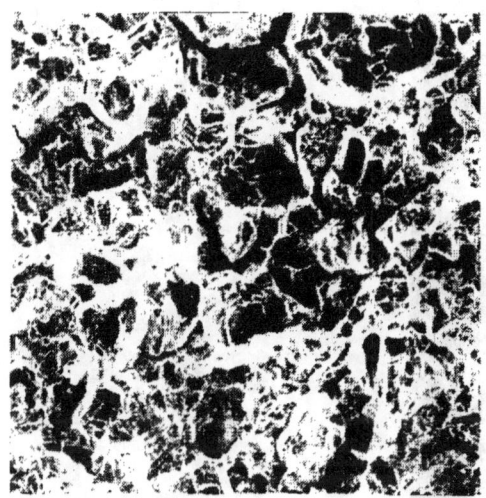

Fig. 4-17   160X
Region A. Central zone of fracture.
Intergranular cracking.

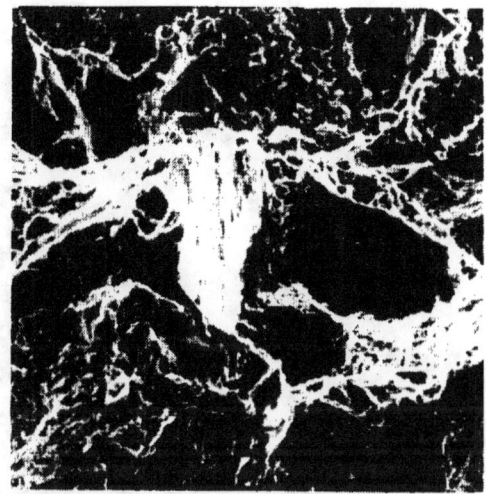

Fig. 4-18   640X
Region A. Enlarged view of center
left of Fig. 4-17.

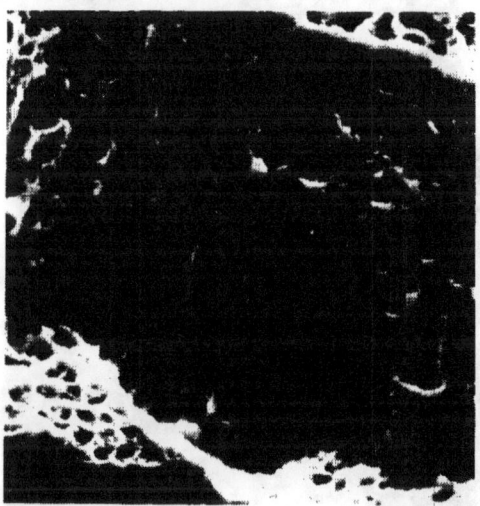

Fig. 4-19   2500X
Region A. Small dimples on the
fracture surface of Fig. 4-18.

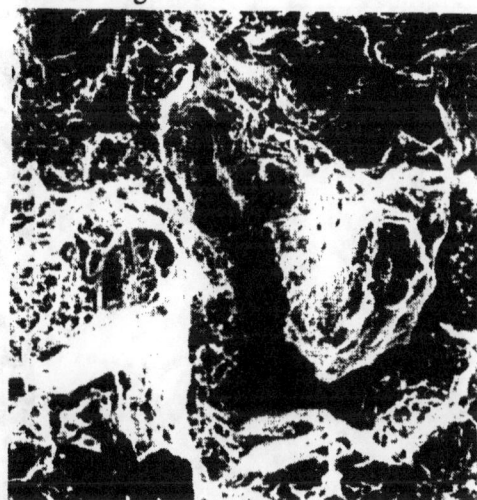

Fig. 4-20   640X
Region A. Enlarged view of the
center of Fig. 4-17.

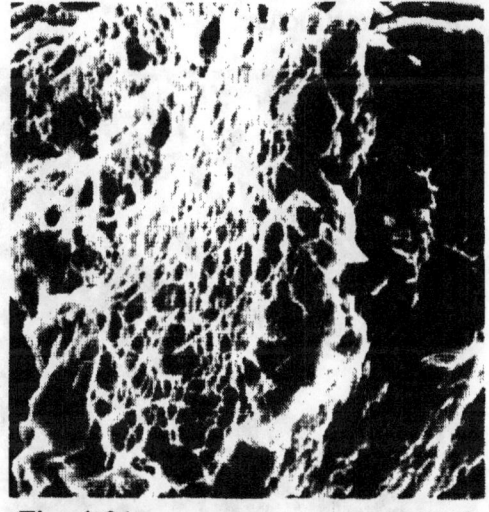

Fig. 4-21   2500X
Region A. Enlarged view of the
microdimples in Fig. 4-20.

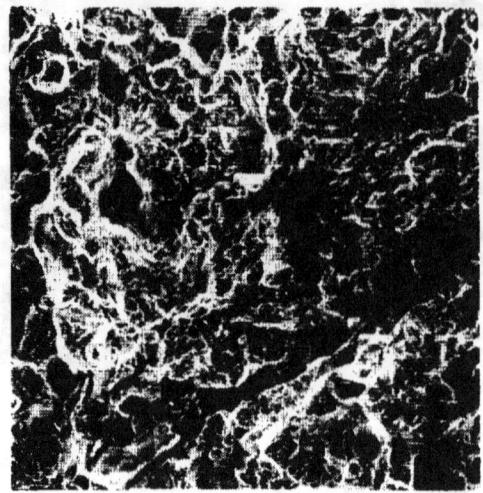

Fig. 4-22   160X
Region B. An area close to the
fracture origin.

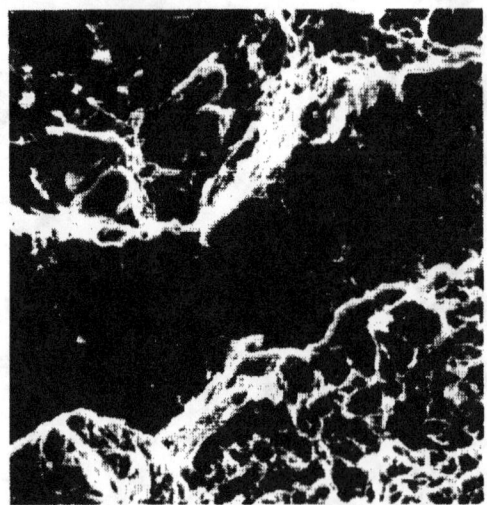

Fig. 4-23   1250X
Region B. Long band associated with
the microstructure.

Fig. 4-24   5000X
Region B. Microdimples.

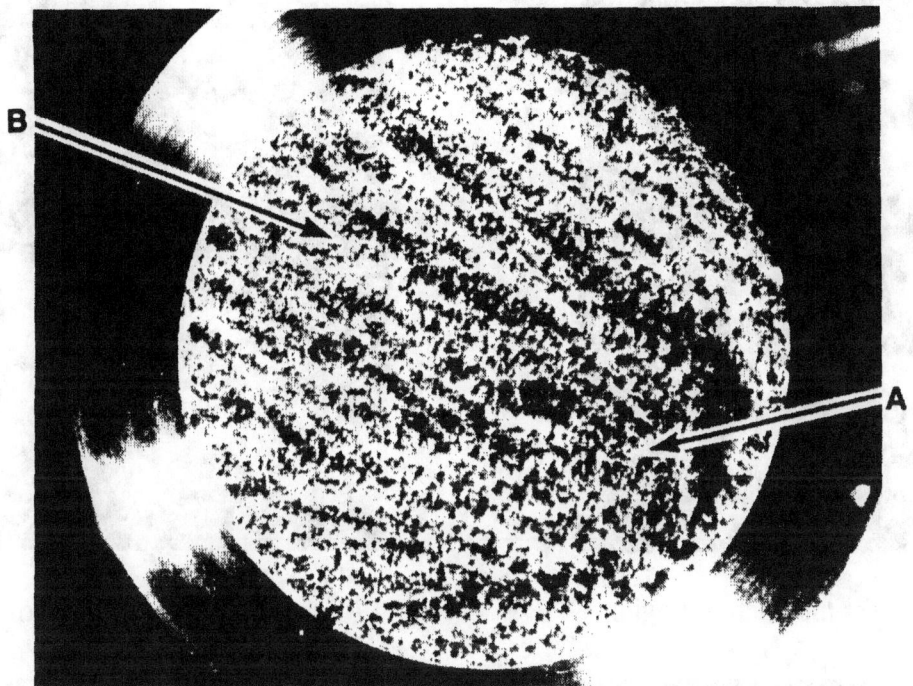

Fig. 4-25   13X
Optical photomacrograph of the fracture surface.

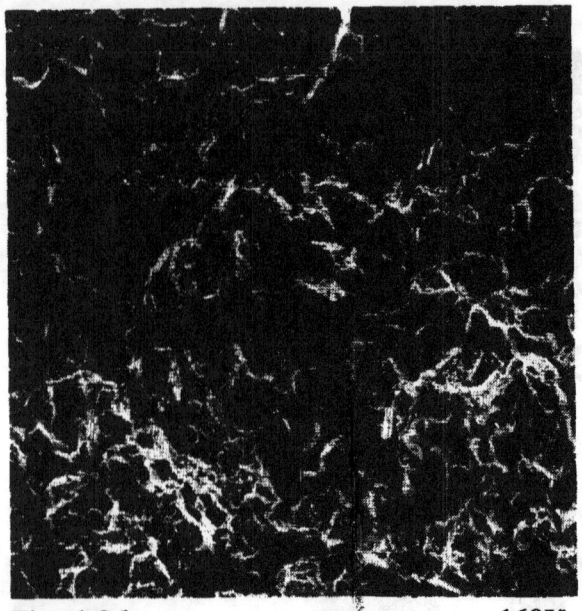

Fig. 4-26   160X
Region A. An area near the fracture origin.

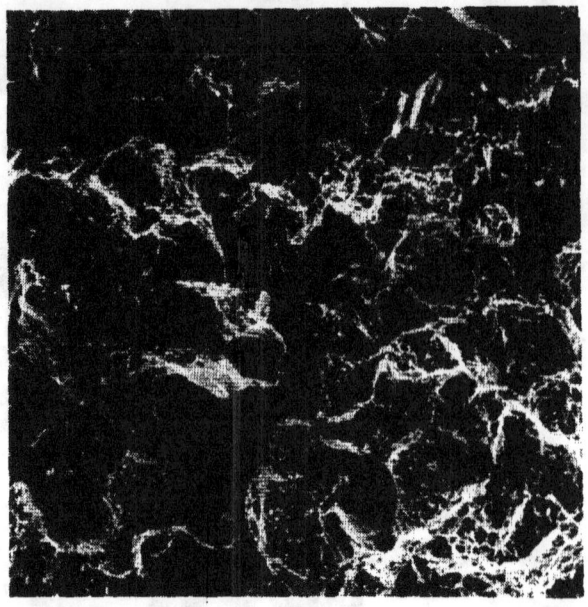

Fig. 4-27   320X
Region A. Enlarged view of Fig. 4-26.
Intergranular cracking.

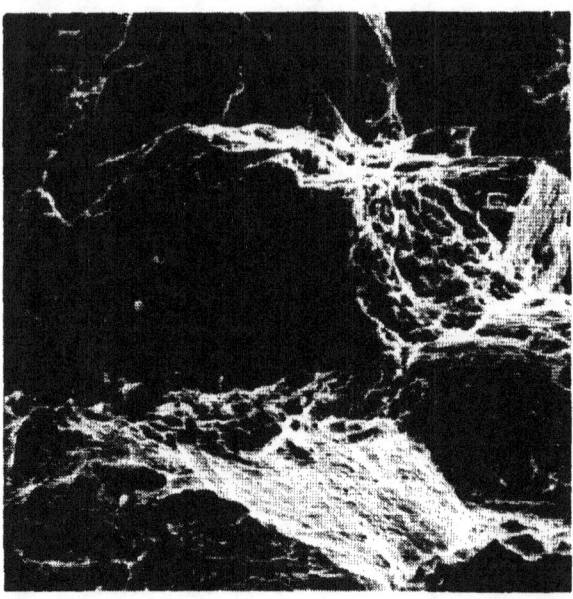

Fig. 4-28   1250X
Region A. Enlarged view of Fig. 4-27. Dimples.

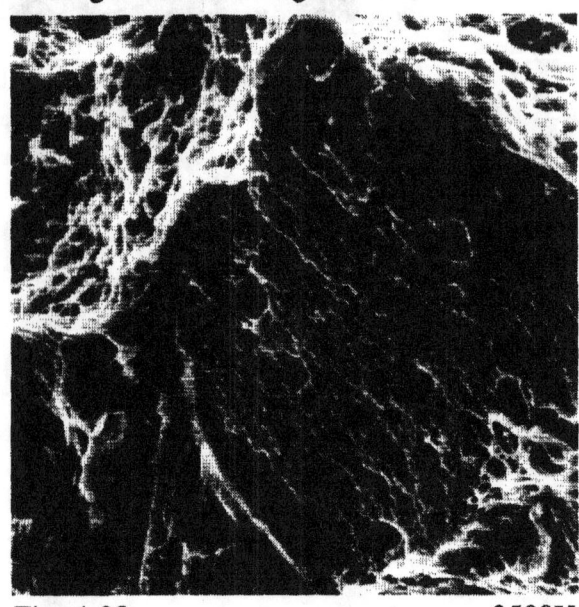

Fig. 4-29   2500X
Region A. Another enlarged view of Fig. 4-27.
Dimples on grain facet.

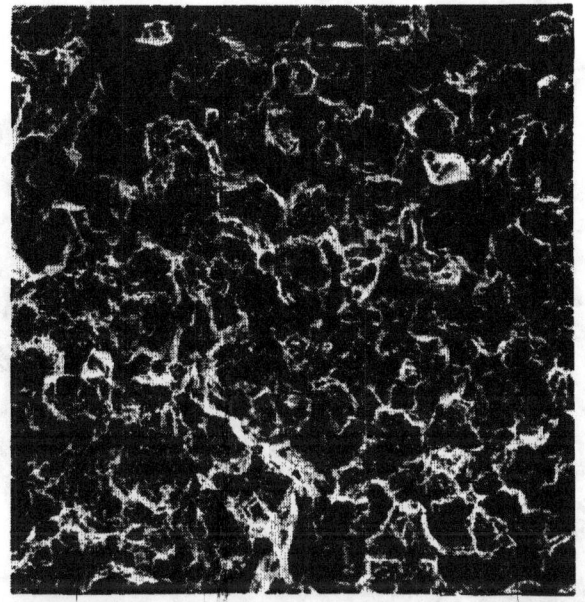

Fig. 4-30   160X
Region B. Area in the central zone of fracture.

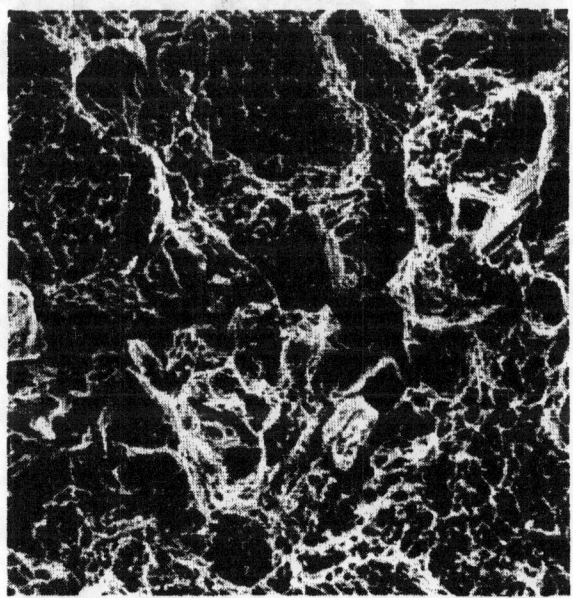

Fig. 4-31   640X
Region B. Area in the center of Fig. 4-30.

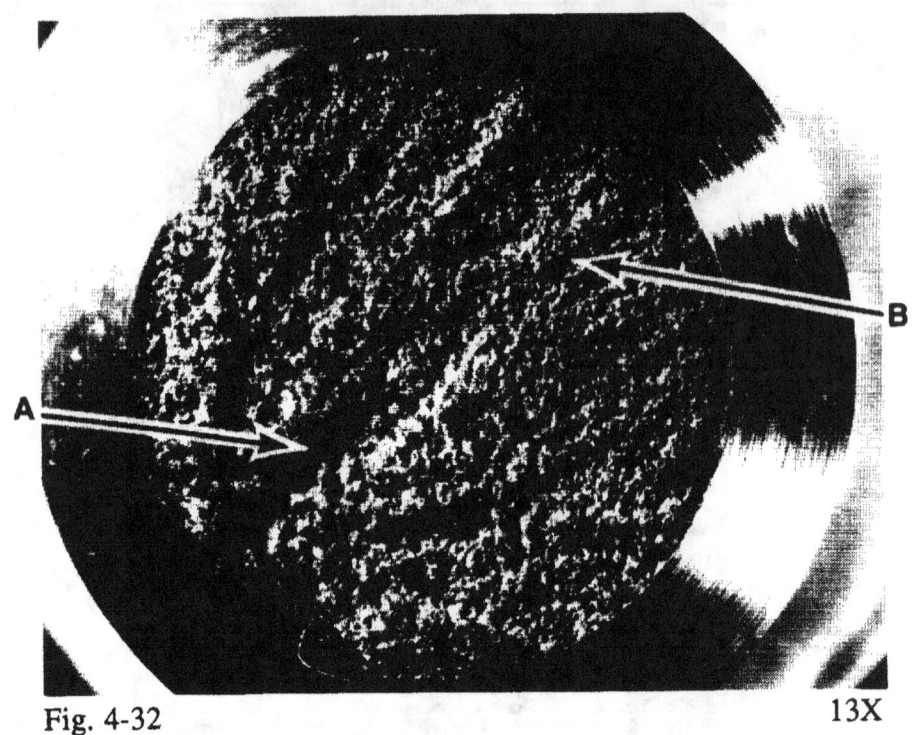

Fig. 4-32   13X

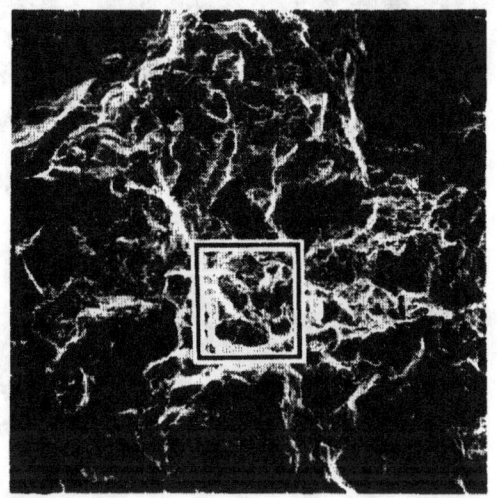

Fig. 4-33   160X
Region A. An area near the
fracture origin.

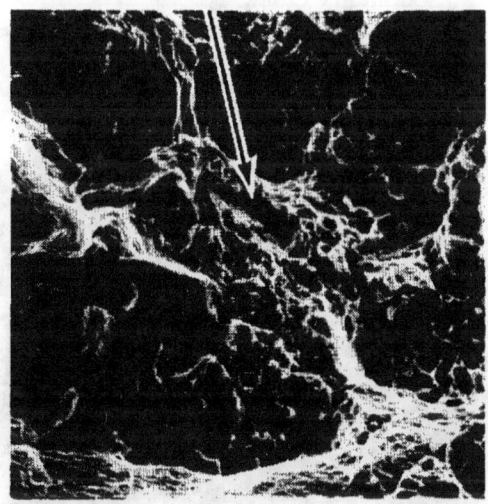

Fig. 4-34   1250X
Region A. Enlarged view of the center
of Fig. 4-33. Arrowhead points to
secondary crack.

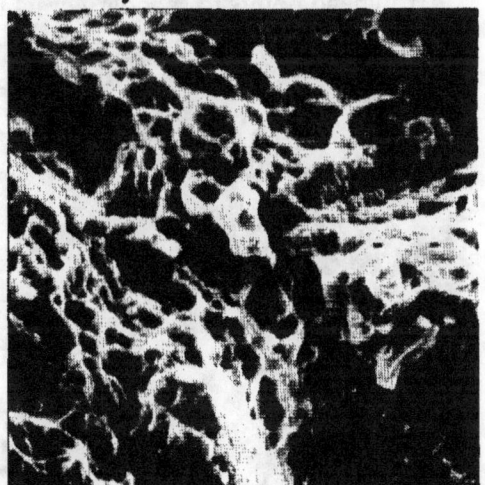

Fig. 4-35   2500X
Region A. Enlarged view of Fig. 4-34.
Microdimples.

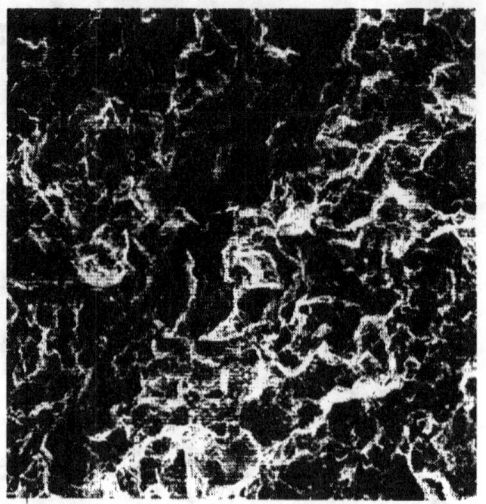

Fig. 4-36   160X
Region B. An area near the center of
the fracture containing a large
secondary crack.

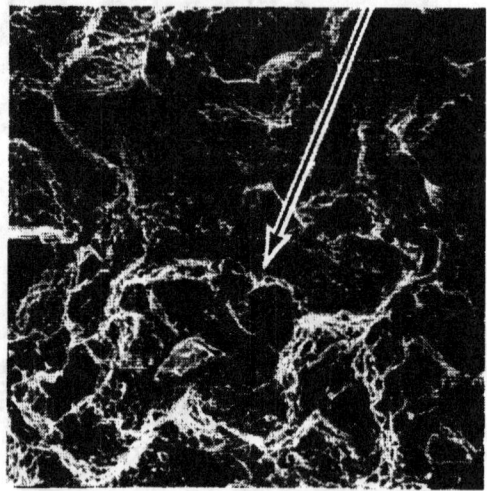

Fig. 4-37   640X
Region B. An area to the right of the
large crack in Fig. 4-36.
Arrowhead points to secodary cracking.

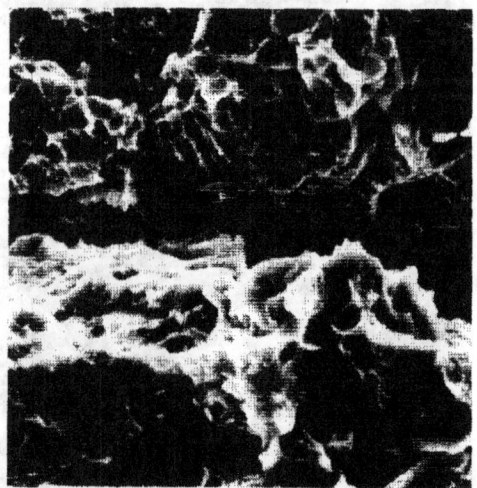

Fig. 4-38   2500X
Region B. Enlarged view of the
secondary crack in Fig. 4-37.

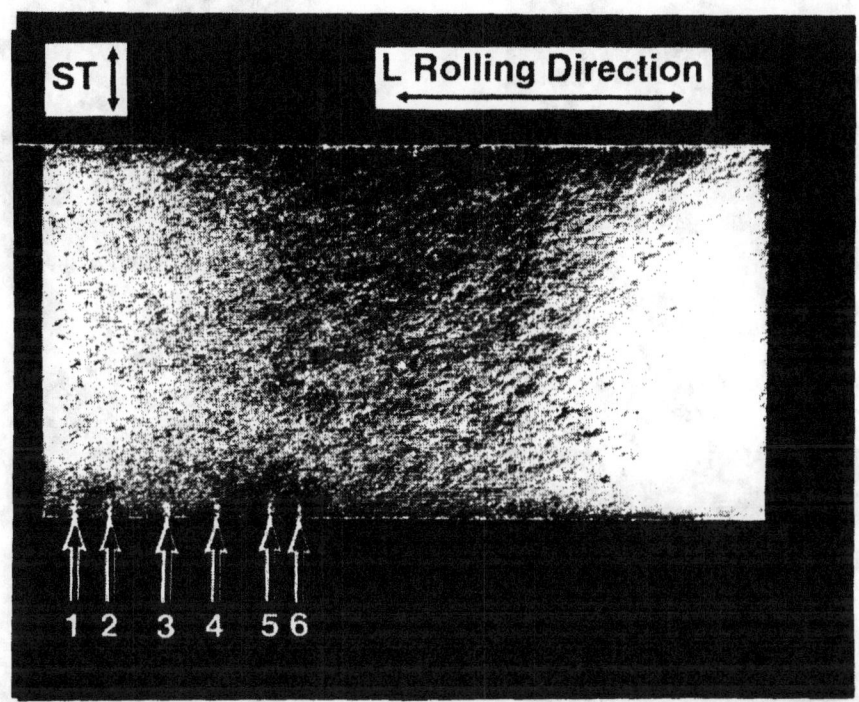

Fig. 4-39   4X
Optical photomacrograph of the fatigue fracture surface.

95

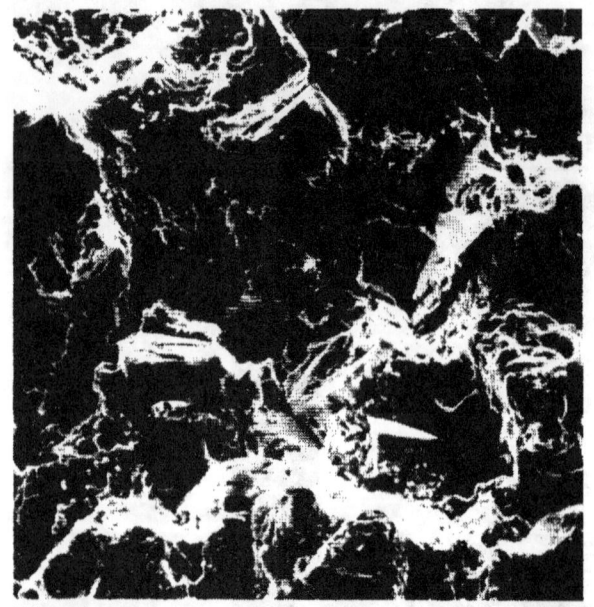

Fig. 4-40   640X
Region 2.

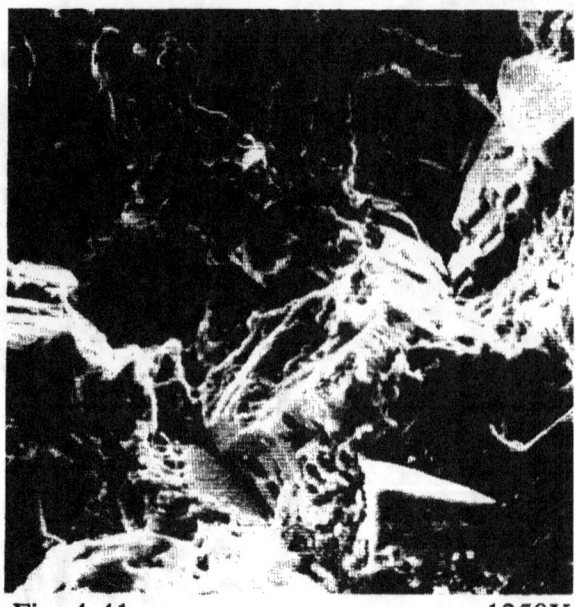

Fig. 4-41   1250X
Region 2.

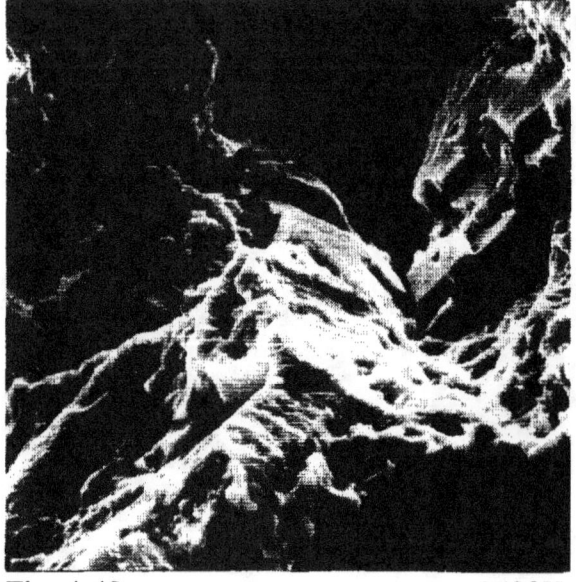

Fig. 4-42   2500X
Region 2.

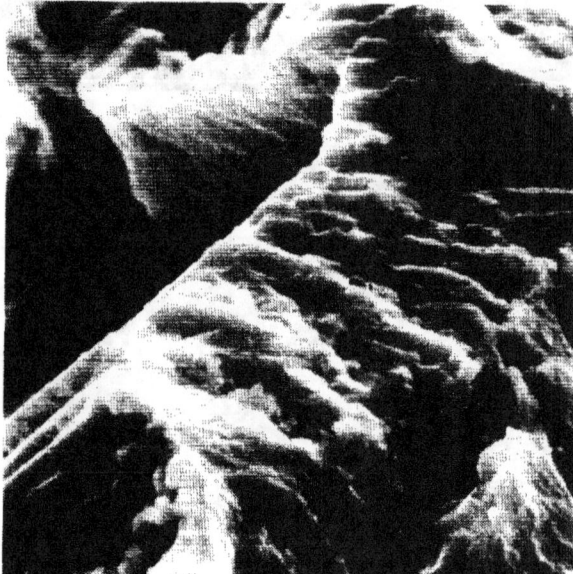

Fig. 4-43   10000X
Region 2.

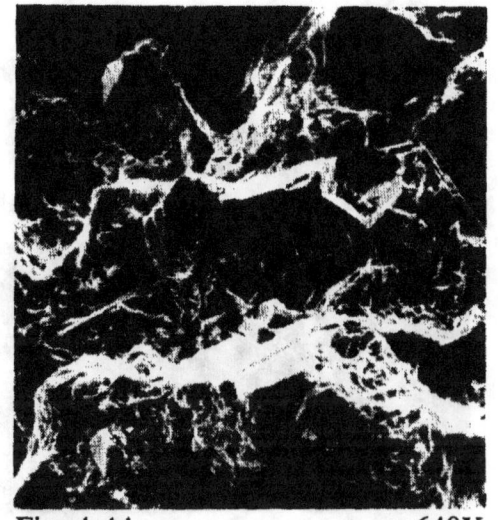

Fig. 4-44   640X
Region 3.

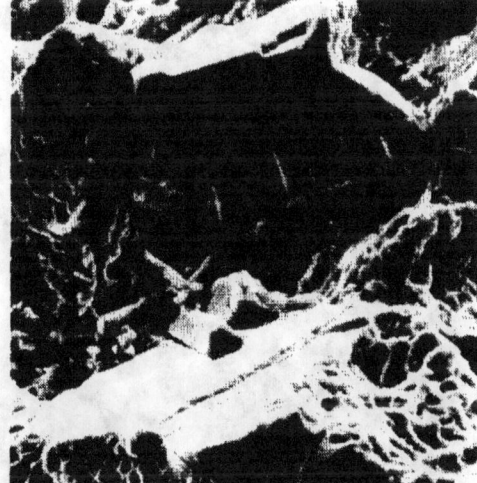

Fig. 4-45   1250X
Region 3.

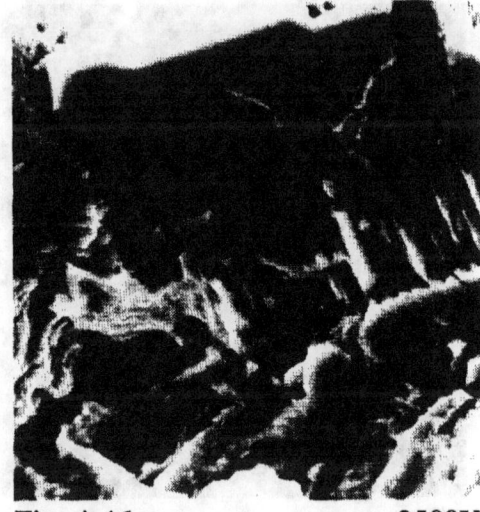

Fig. 4-46   2500X
Region 3.

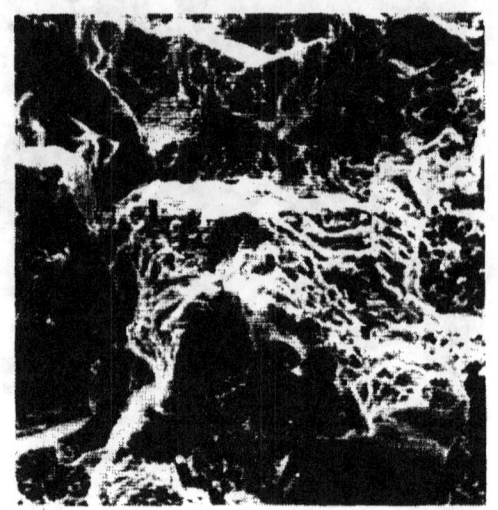

Fig. 4-47   640X
Region 4.

Fig. 4-48   2500X
Region 4.

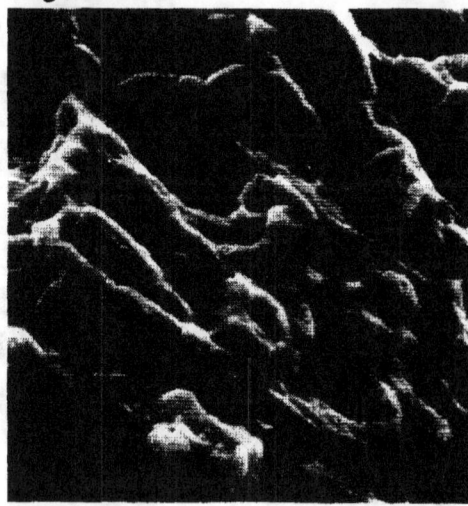

Fig. 4-49   5000X
Region 4.

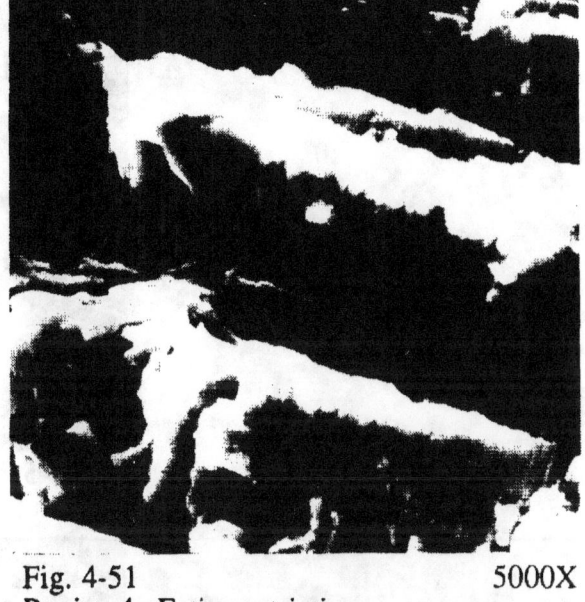

Fig. 4-50   640X
Region 4.

Fig. 4-51   5000X
Region 4. Fatigue striations.

Fig. 4-52   10,000X
Region 4. Fatigue striations.

Fig. 4-53   20000X
Region 4. Fatigue striations.

99

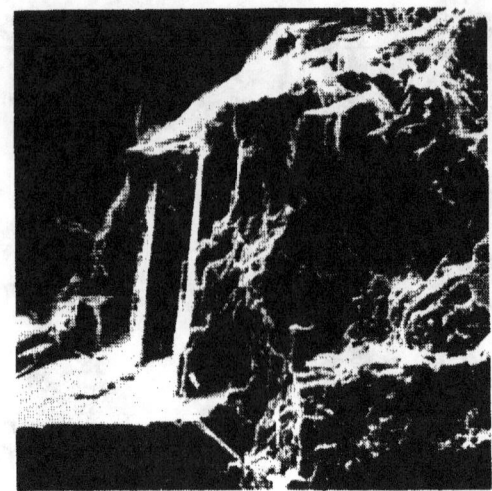

Fig. 4-54   640X
Region 5.

Fig. 4-55   2500X
Region 5.

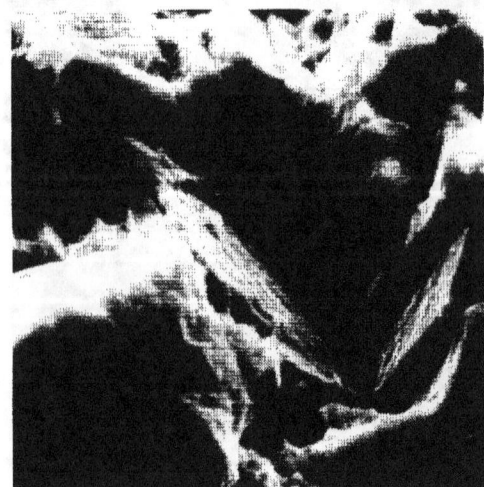

Fig. 4-56   5000X
Region 5.

Specimen 5 - 6061 Aluminum Welded With 4043 Aluminum

Specimen Condition:
Two pieces of 6061 aluminum were welded together with 4043 filler metal as shown in Figure 5-a.
Dimensions: 0.5" Tk X 4" W X 16" L. The specimen did not weld properly and thus there is an
area where there is lack of penetration. The lack of penetration is evident in the microstructure as
well as in the fracture surfaces.

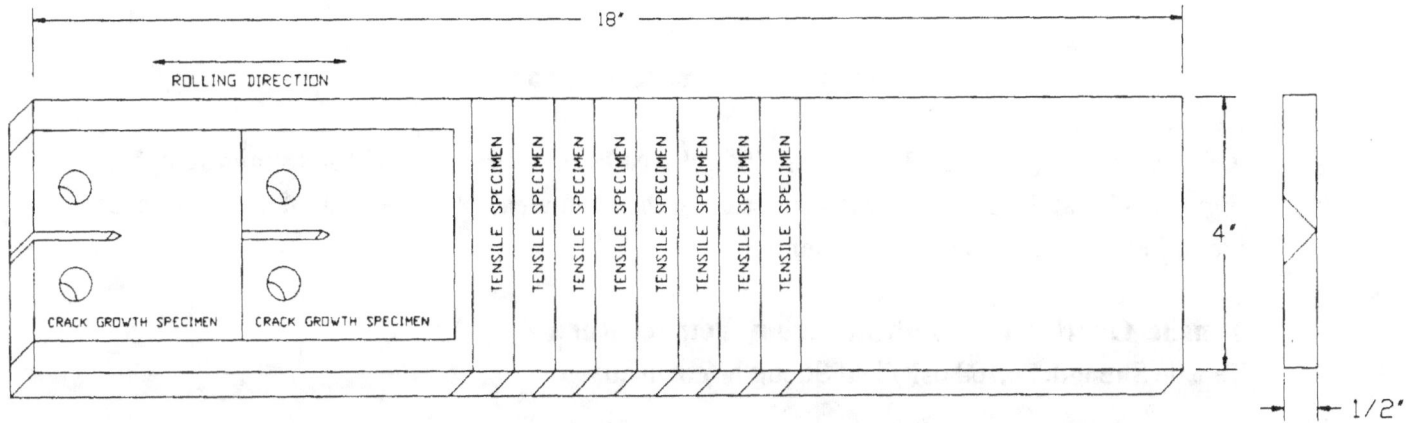

Figure 5-a. Diagram of pattern used to cut welded test specimens

Specimen Geometry:
The dimensions for the smooth tensile overload specimens are shown in Figure 5-b, and the
dimensions for the notched tensile overload specimens are shown in Figure 5-c. The compact
tension specimen dimensions are shown in Figure 5-d.

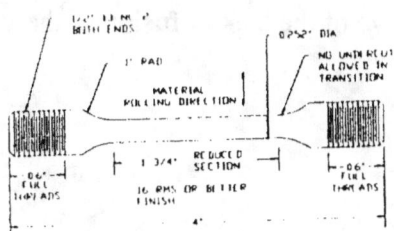

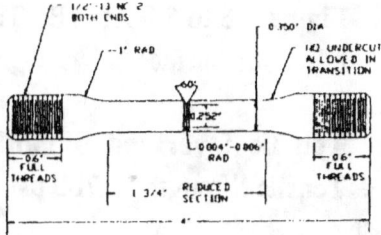

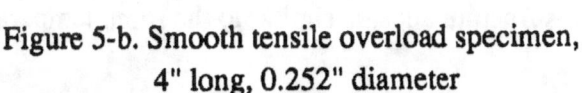

Figure 5-b. Smooth tensile overload specimen, 4" long, 0.252" diameter

Figure 5-c. Notched tensile overload specimen, 4" long, 0.004-0.006" root radius, 0.252" notch diameter.

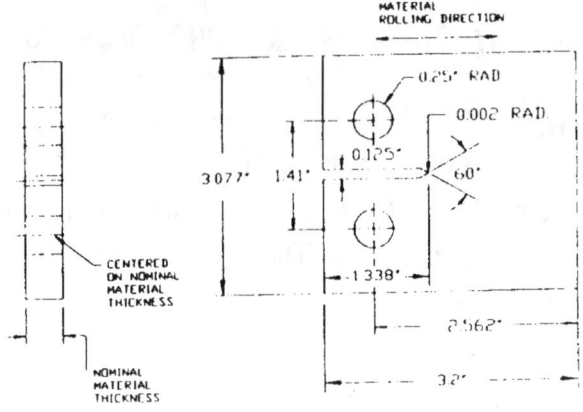

Figure 5-d. Compact tension specimen

Metallography:

Keller's etchant (1 ml HF, 1.5 ml HCl, 2.5 ml HNO_3, 95 ml H_2O) used for all micrographs. Figures 5-1 and 5-2 show the microstructure of the weld area. Note the lack of penetration area identified with the arrowhead in Figure 5-1.

Tensile Overload Smooth at Room Temperature

Tensile Strength 24,100 psi Yield Strength- not available

<u>Macroscopic Appearance</u>

The fracture surface is shown in Figure 5-3. The surface is rough and uneven. The smooth area across the center of the specimen is a result of incomplete penetration of the weld and is not a fracture surface. The central portion is the nonwelded area.

<u>Microscopic Appearance</u>

Figures 5-4 to 5-7 Area A: This area is located on the left side of the fracture as shown in Figure 5-3. These microfractographs are representative of the welded metal. The fracture shows ductility at both low magnification and high magnification.

Figure 5-8 to 5-9 Area B: This area is representative of the lack of fusion of the weld which is not a fracture surface.

Tensile Overload Smooth at 77 K

Tensile Strength 25,700 psi

<u>Macroscopic Appearance</u>

The fractograph in Figure 5-10 shows that this fracture appears similar to the room temperature sample.

<u>Microscopic Appearance</u>
Figure 5-11 shows a SEM micrograph of the surface.

Figure 5-12 to 5-15 Area A: These micrographs show dendrites in a porous weld area. This is the surface of a void area and is not a fracture surface.

Figure 5-16 to 5-17 Area B: This area represents the incomplete penetration of the weld.

Figure 5-18 to 5-19 Area C: A region of ductile dimples.

Tensile Overload Notched Room Temperature
Tensile Strength 30,700 psi

<u>Macroscopic Appearance</u>
Figure 5-20 shows the macrograph of this fracture surface.

<u>Microscopic Appearance</u>
Figure 5-21 shows a low magnification SEM micrograph of the fracture surface.

Figure 5-22 to 5-23 Area A: Fractographs showing the rough, dimpled fracture typical of this region.

Figure 5-24 to 5-25 Area B: Fractographs showing the smooth nearly featureless unwelded area.

Figure 5-26 to 5-27 Area C: Fractographs showing the equiaxed dimples typical of this region.

Tensile Overload of Notched Sample at 77 K.
Tensile Strength 28,800 psi

<u>Macroscopic Appearance</u>

An optical picture of the fracture is shown in Figure 5-28.

Figure 5-29 shows a SEM micrograph of the fracture.

Figure 5-30 to 5-33 Area A: This area shows a rough and dimpled fracture surface in Figures 5-30 to 5-31. Dendritic formation in void areas is also evident in this area as shown in Figures 5-32 and 5-33.

Figure 5-34 to 5-35 Area B: These micrographs show the smooth features of the non-welded area resulting from incomplete penetration.

HIGH CYCLE FATIGUE TEST DATA

<u>Macroscopic Appearance</u>
Two of the fatigue regions are identified in the optical photograph in Figure 5-36. Regions 2 and 3 were documented using the SEM. Regions 1 and 4 are the precrack and fast fracture regions respectively and were not documented with fractographs. The loading data for these regions is displayed in the following chart:

Region	R	K_{max} (ksi-in$^{1/2}$)	dA/dN
2	0.5	10	not available
3	0.75	20	not available

<u>Microscopic Appearance</u>
Figures 5-37 to 5-39 Region 2: The striations are clearly evident under these load conditions. Note the inclusions in the fracture surface.

Figures 5-40 to 5-42 Region 3: The striations are evident at this load ratio but the striation spacing is smaller due to the higher R value.

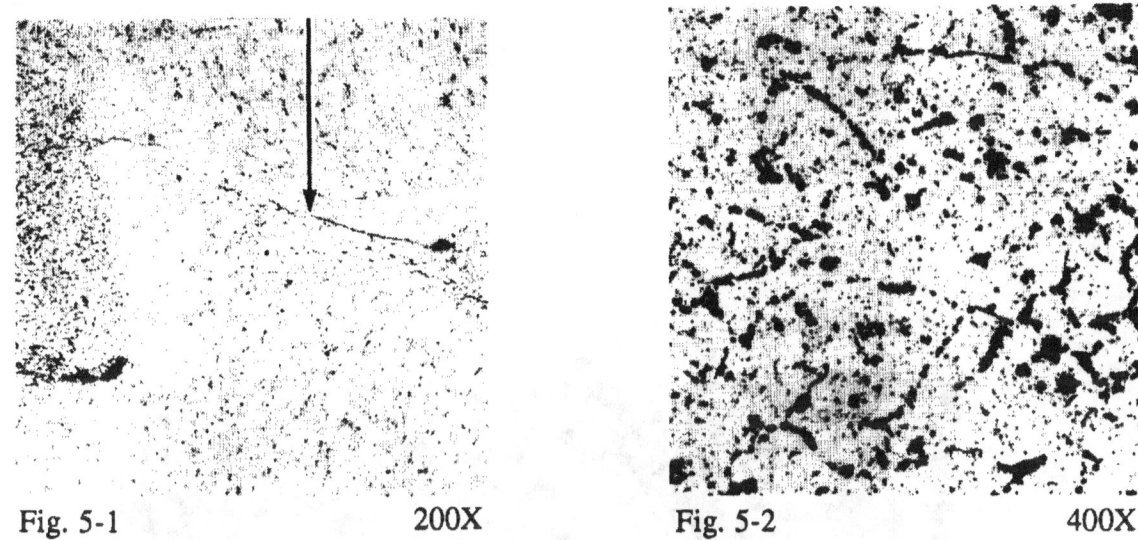

Fig. 5-1   200X   Fig. 5-2   400X

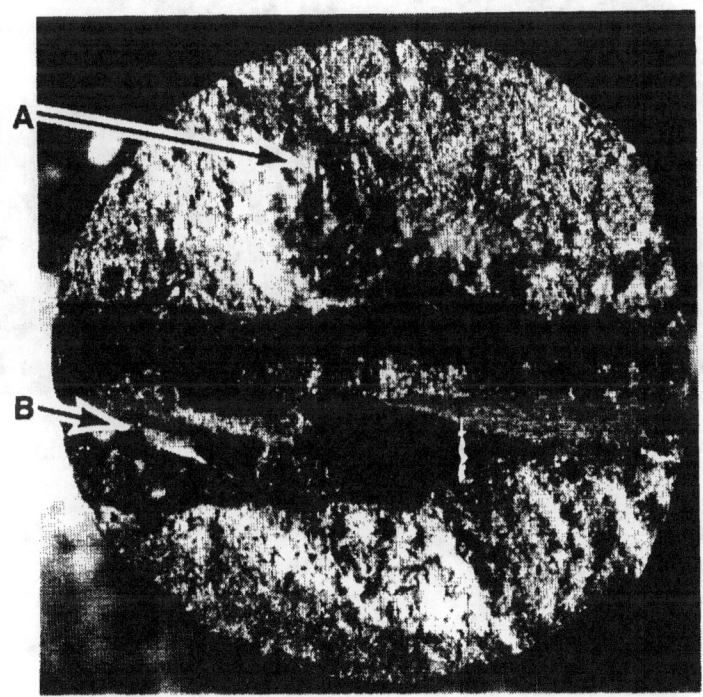

Fig. 5-3   14X
Optical photomacrograph of fracture surface.

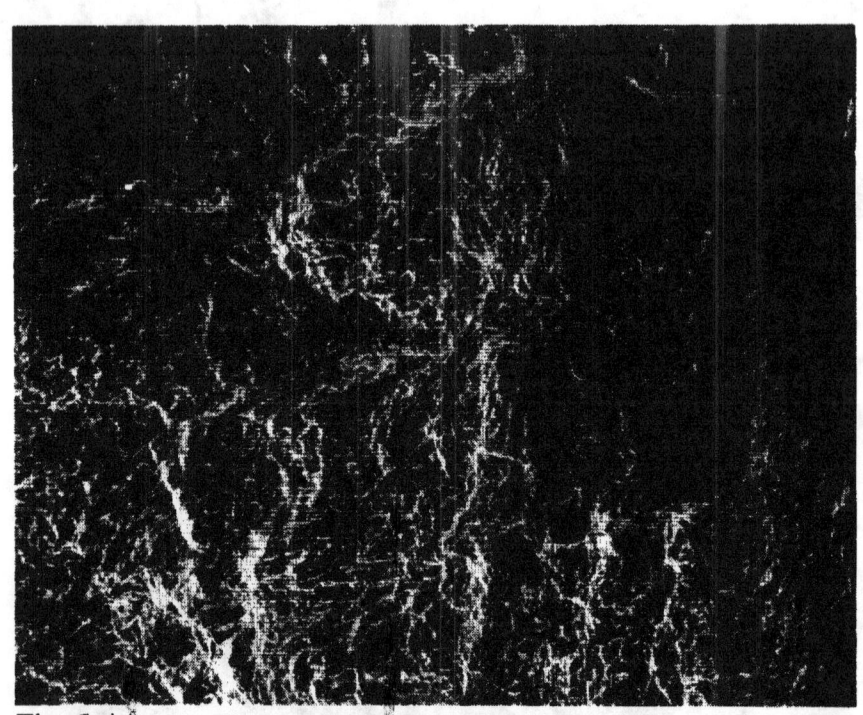

Fig. 5-4   40X
Low magnification view of the different fracture morphologies.

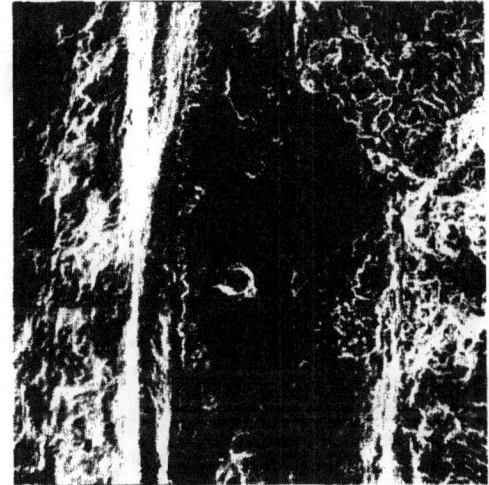

Fig. 5-5   160X
Region A. Ductile fracture.

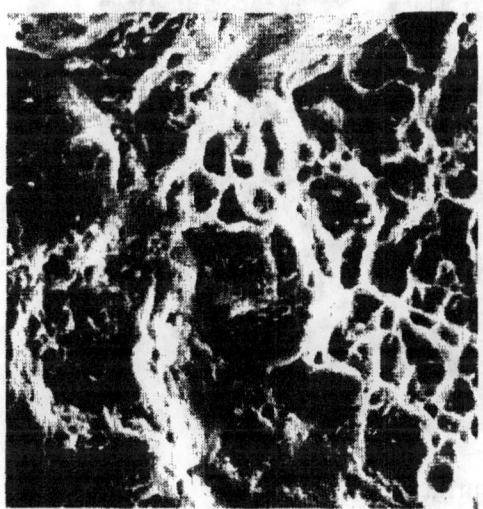

Fig. 5-6   640X
Region A. Enlarged view of an area
in the center of Fig. 5-5.

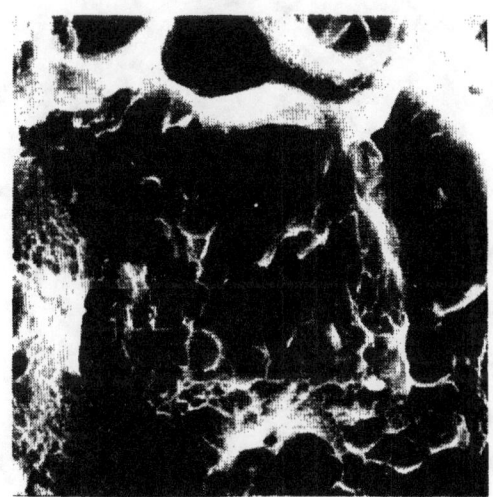

Fig. 5-7   2500X
Region A. Center of Fig. 5-7.

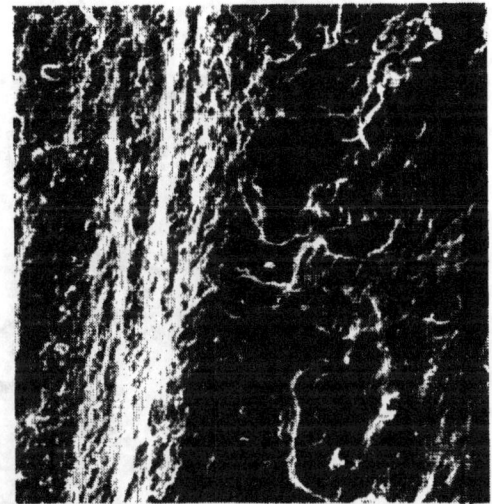

Fig. 5-8   160X
Region B. Area showing lack of
penetration.

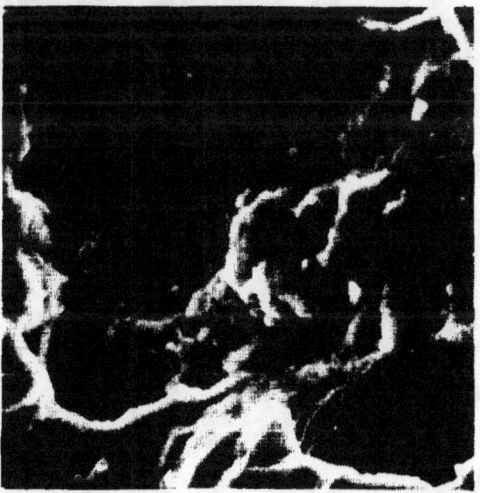

Fig. 5-9   640X
Region B. Center of Fig. 5-8.

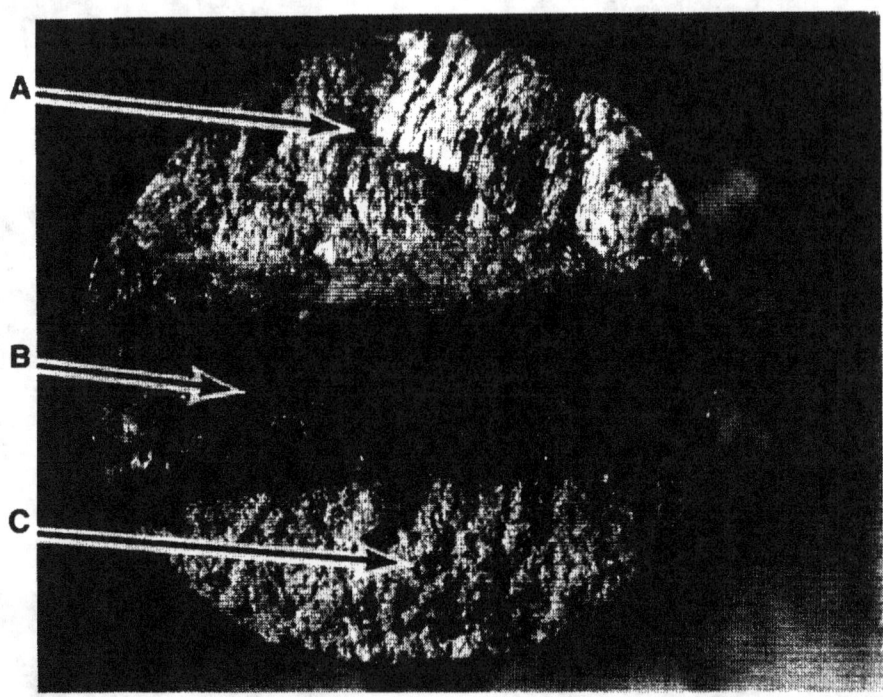

Fig. 5-10   14X
Optical photomacrograph of fracture surface.

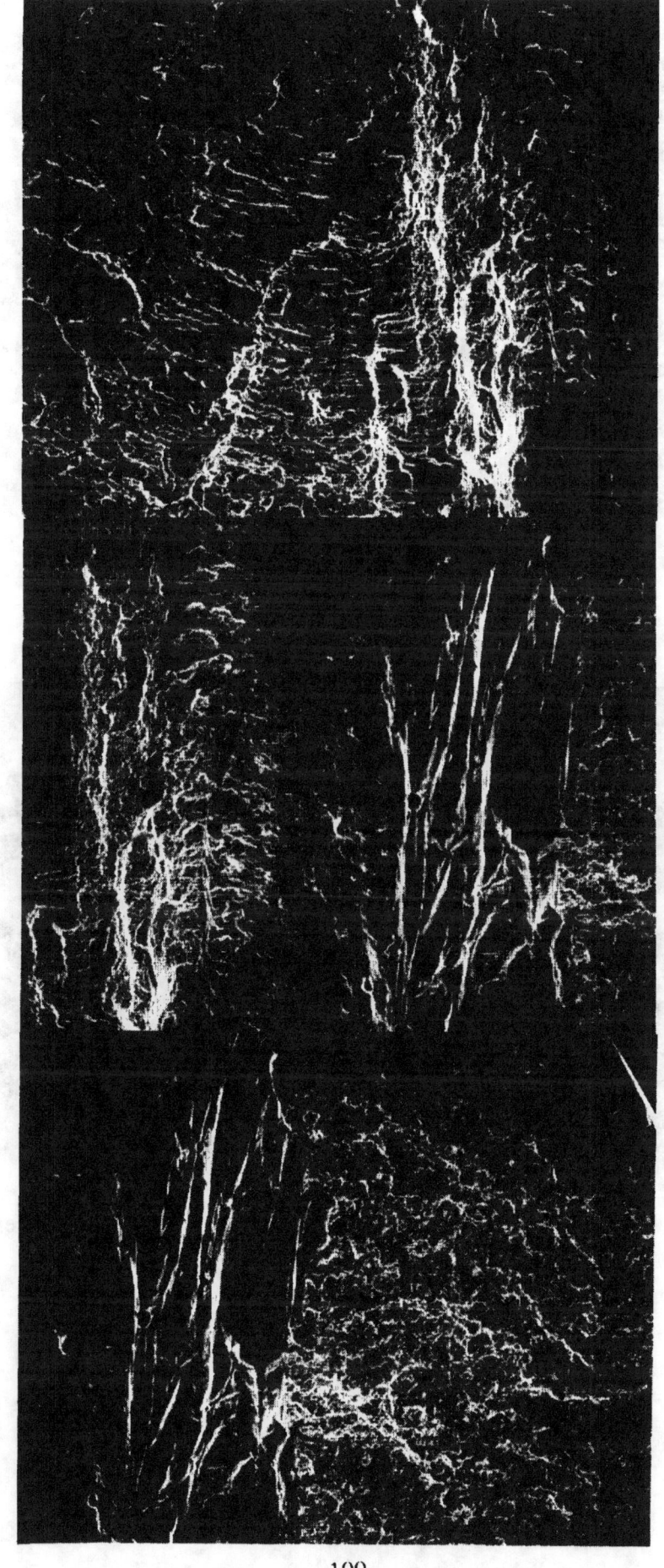

40X

Fig. 5-11
Low magnification fractograph showing several fracture morphologies.

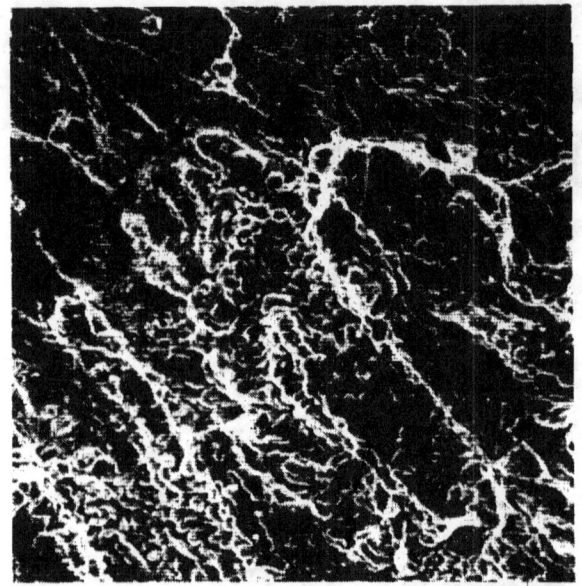

Fig. 5-12   160X
Region A. Dendrites

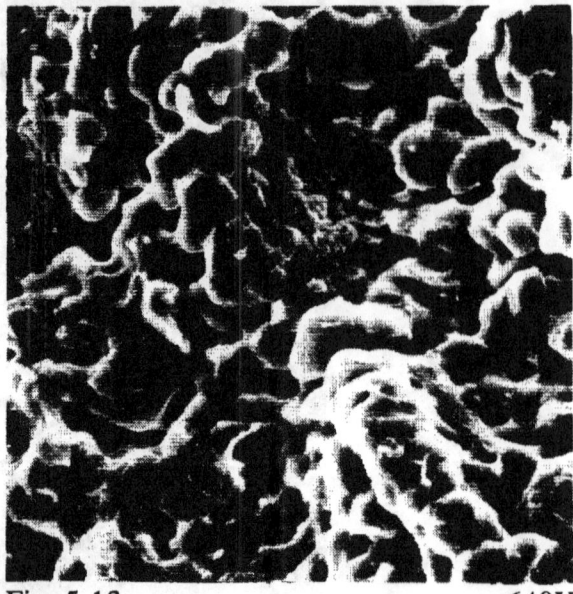

Fig. 5-13   640X
Region A. Enlarged view of the center area
in Fig. 5-12.

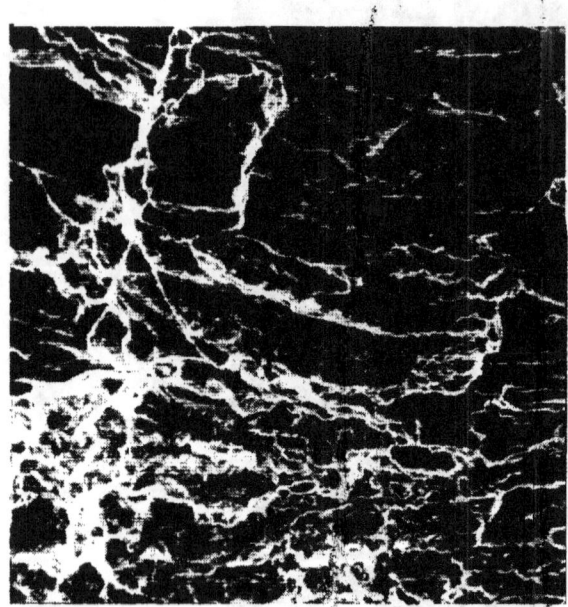

Fig. 5-14   160X
Region A. Another area with dendrites.

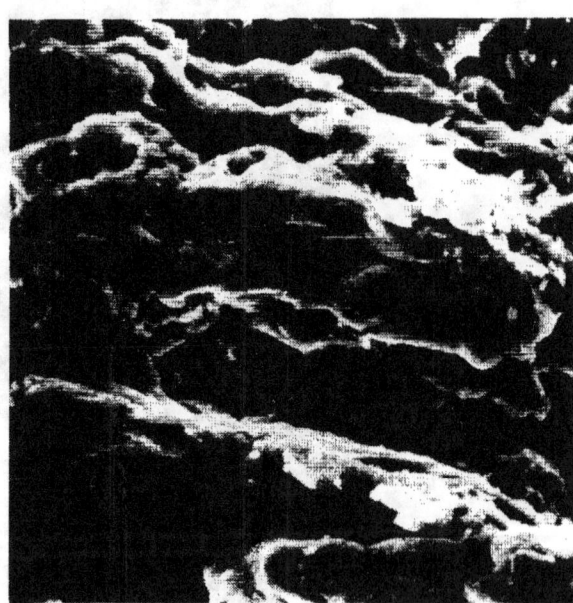

Fig. 5-15   640X
Region A. An area in the center of Fig. 5-14.

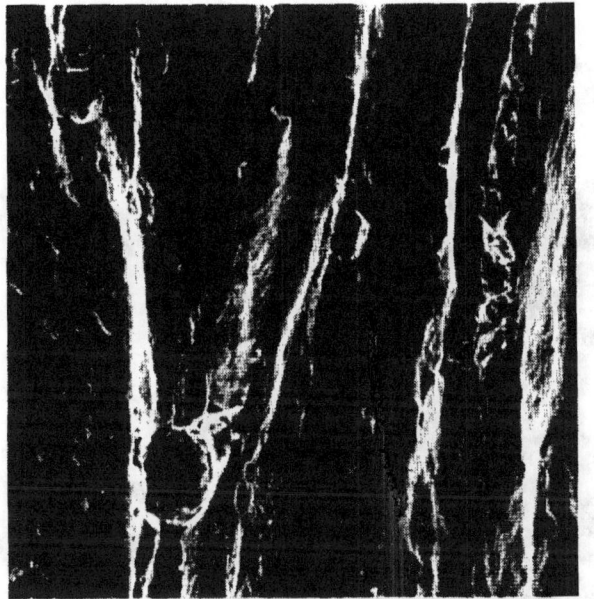

Fig. 5-16   160X
Region B. Lack of penetration of the weld.

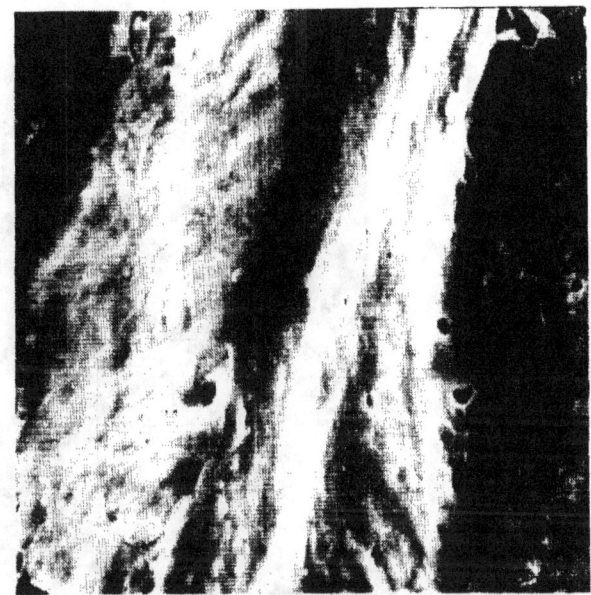

Fig. 5-17   640X
Region B. Enlarged view of center of
Fig. 5-16.

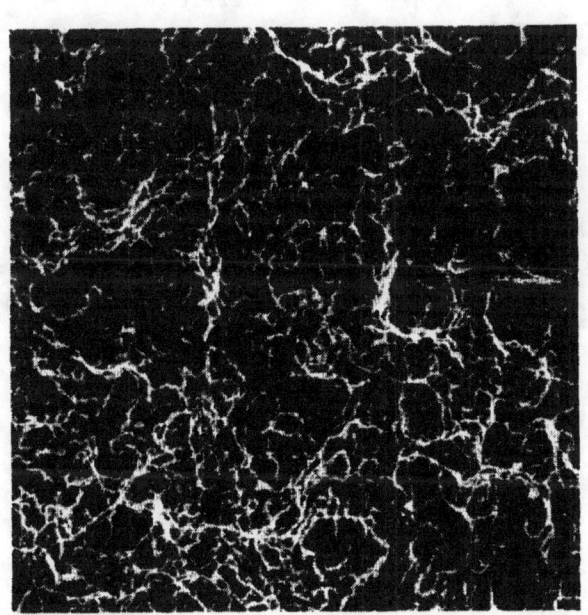

Fig. 5-18   160X
Region C. Ductile dimples.

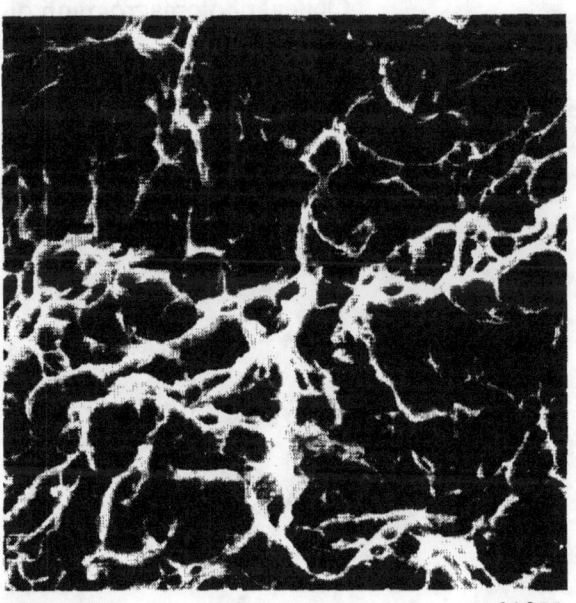

Fig. 5-19   640X
Region C. Enlarged view of the center of
Fig. 5-18.

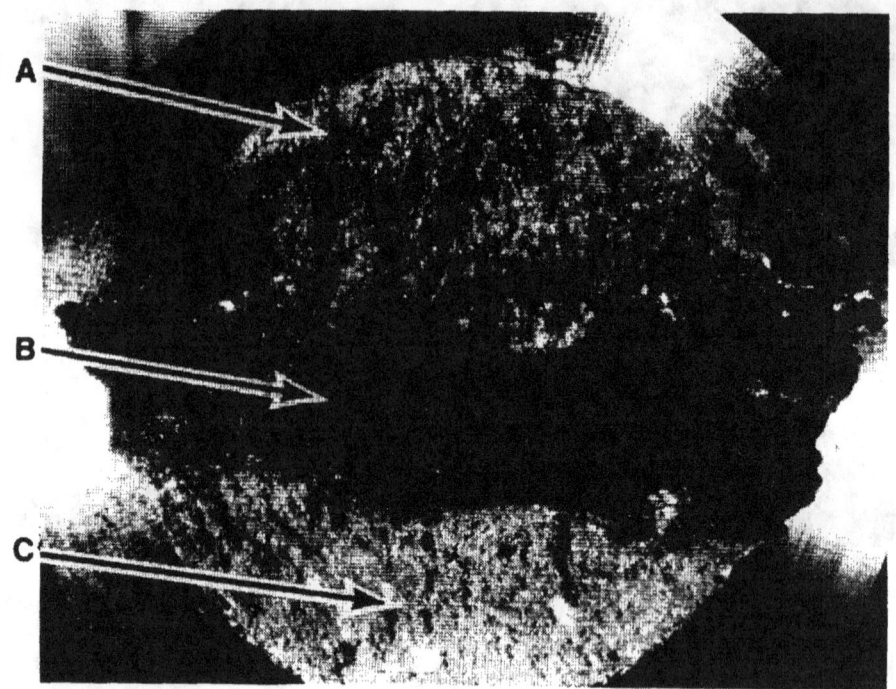

Fig. 5-20   14X
Optical photomacrograph of fracture surface.

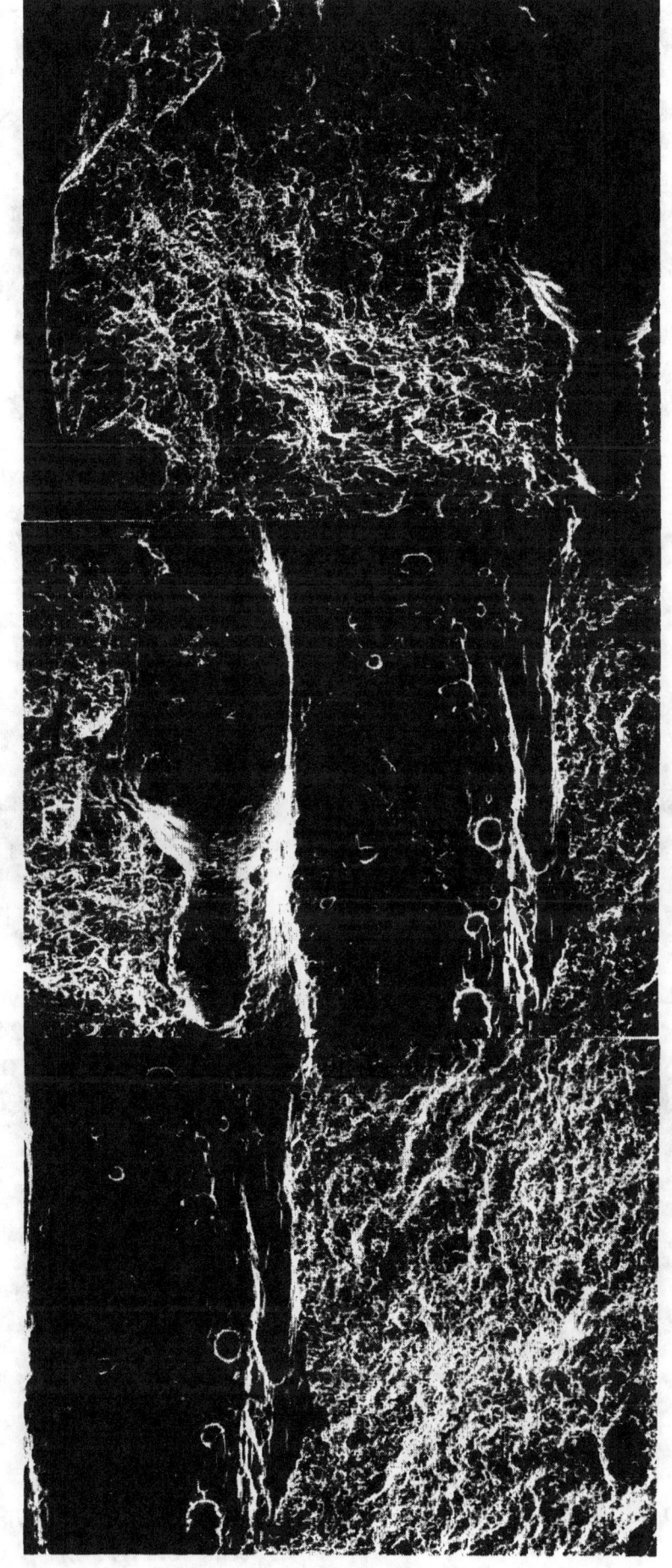

40X

Fig. 5-21
Low magnification fractograph showing several fracture morphologies.

Fig. 5-22   160X
Region A. Ductile dimples.

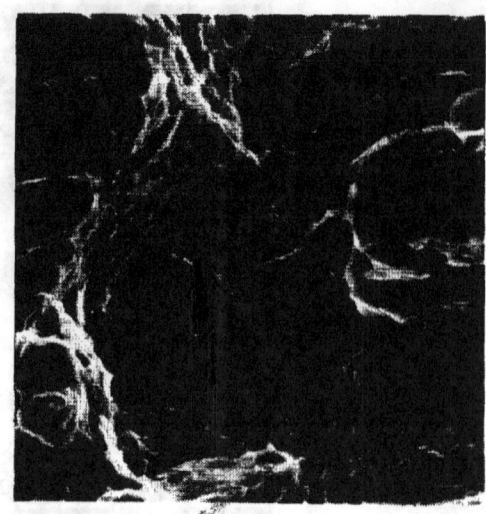

Fig. 5-23   640X
Region A. Enlarged view of the
center of Fig. 5-22.

Fig. 5-24   160X
Region B. Lack of penetration of
the weld.

Fig. 5-25   640X
Region B. Enlarged view of the
boxed area in Fig. 5-24.

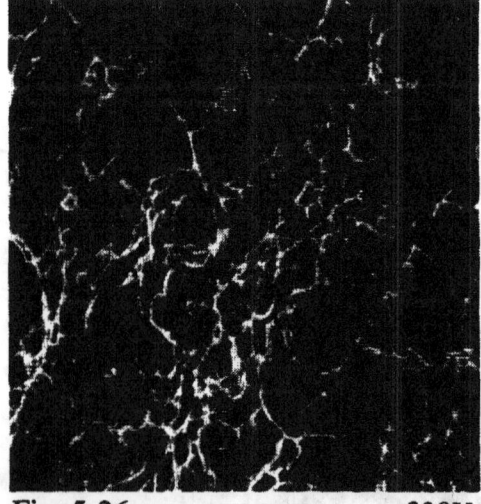

Fig. 5-26   320X
Region C. Ductile dimples.

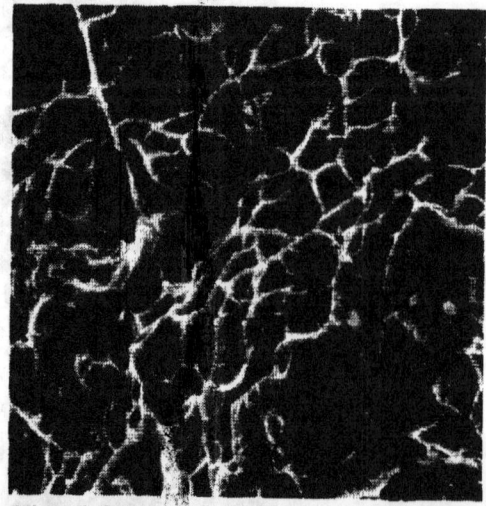

Fig. 5-27   640X
Region C. Enlarged view of the
center of Fig. 5-27.

114

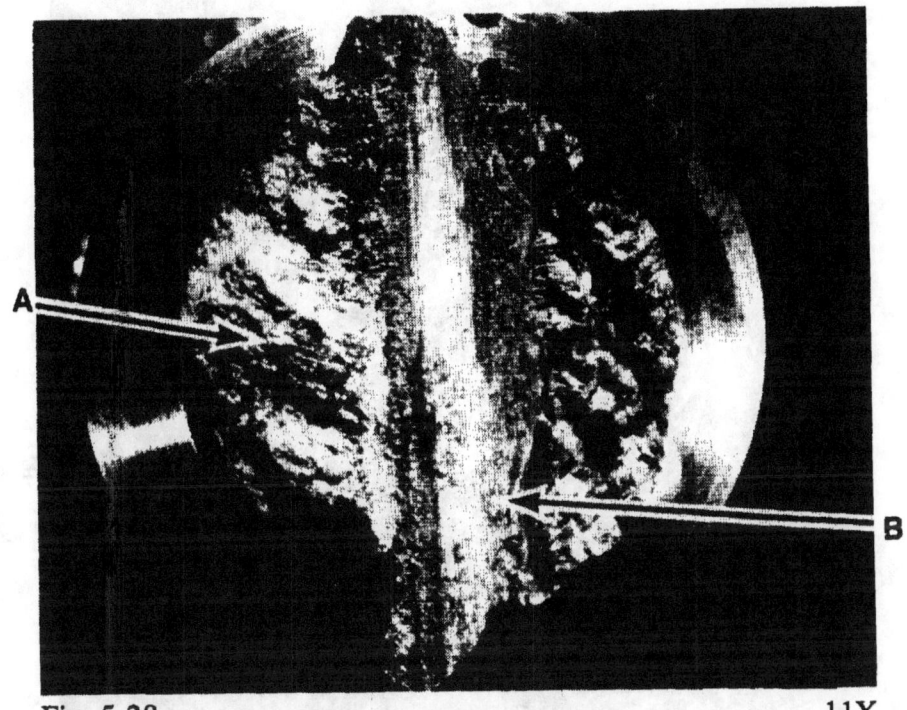

Fig. 5-28   11X
Optical photomacrograph of the fracture surface.

Fig. 5-29
Low magnification fractograph of the different fracture morphologies.

40X

116

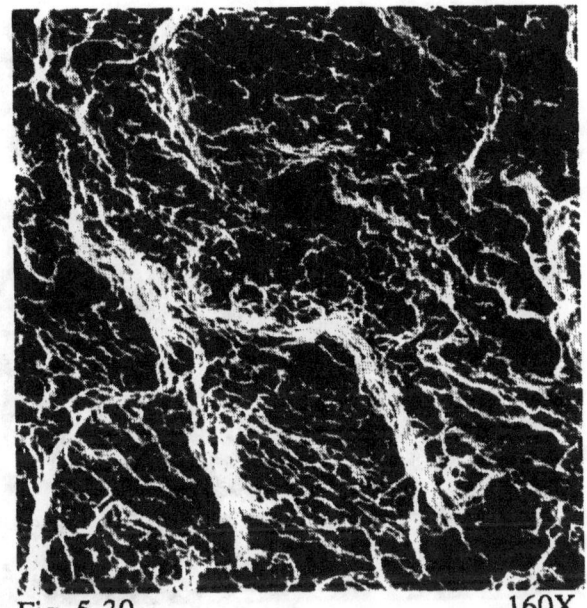

Fig. 5-30   160X
Region A. Ductile dimples.

Fig. 5-31   1250X
Region A. Enlarged view of the center of
Fig. 5-30.

Fig. 5-32   160X
Region A. Area to the right of the dimples in
area A. Dendrites.

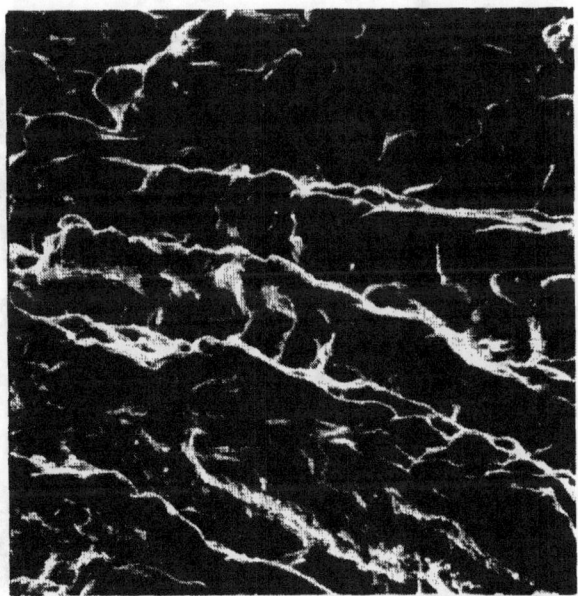

Fig 5-33   320X
Region A. Enlarged view of the dendrites
in Fig. 5-32.

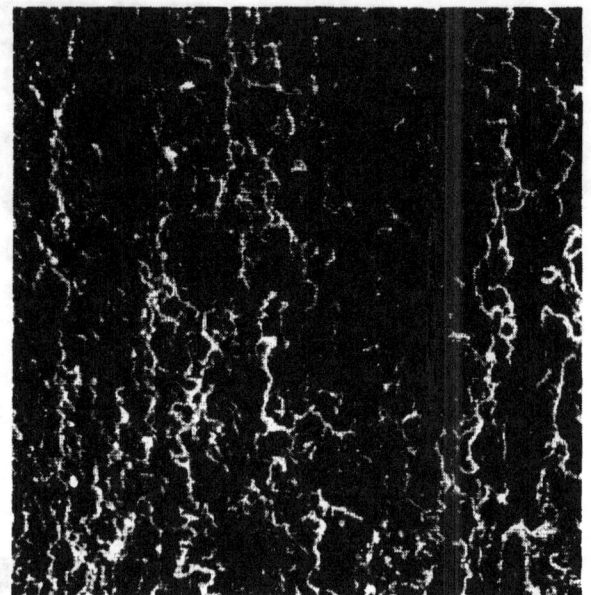

Fig 5-34   160X
Region B. Lack of penetration of the weld.

Fig. 5-35   640X
Region B. Enlarged view of the center of
Fig. 5-34.

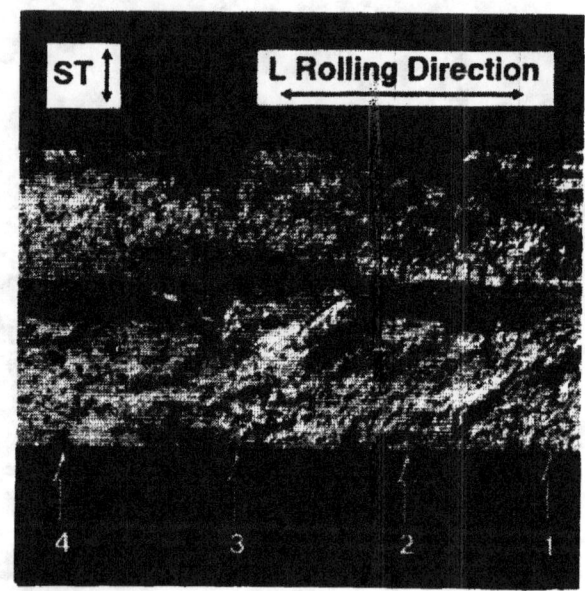

Fig. 5-36   160X
Region C. Dendrites.

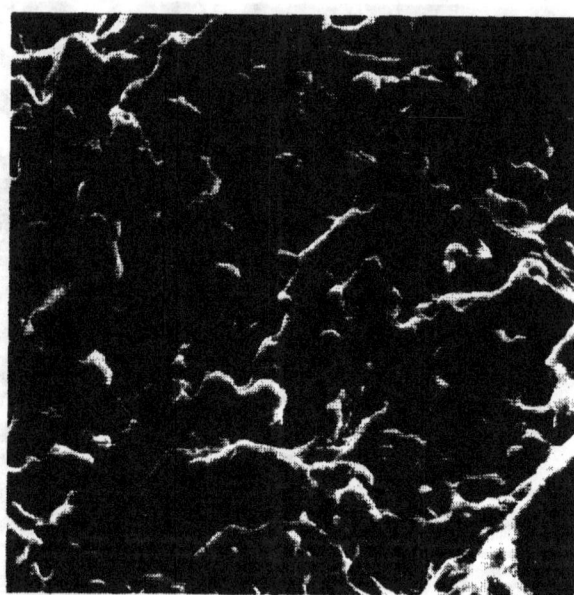

Fig. 5-37   320X
Region C. Enlarged view of the center of
Fig. 5-36.

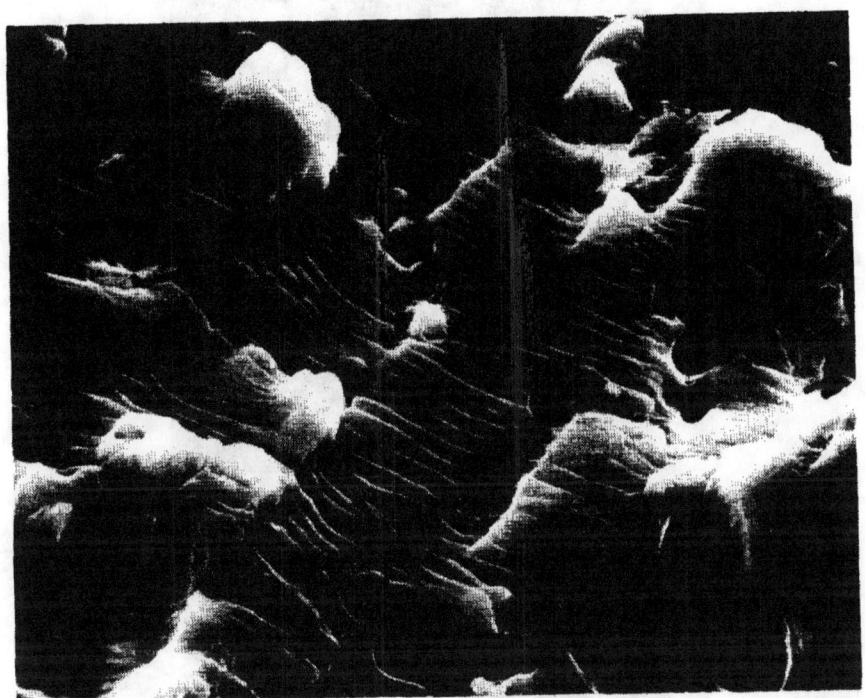

Fig. 5-38   4X
Optical photomacrograph of fatigue fracture surface.

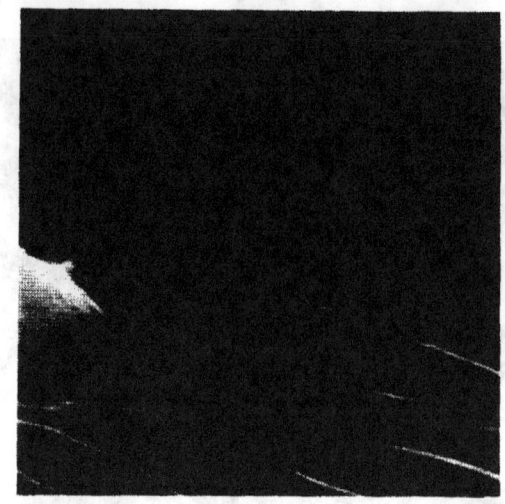

Fig. 5-39   640X
Region 2. Fatigue striations.

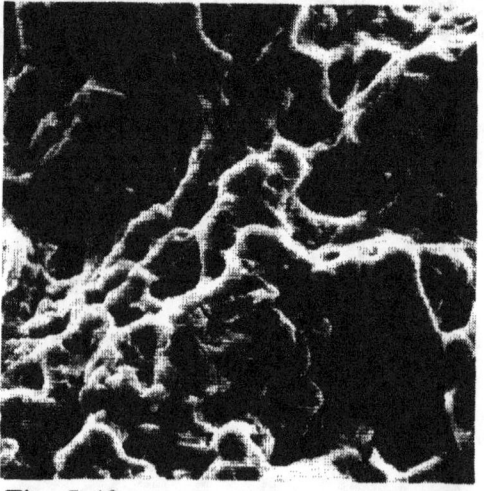

Fig. 5-40   2500X
Region 2.

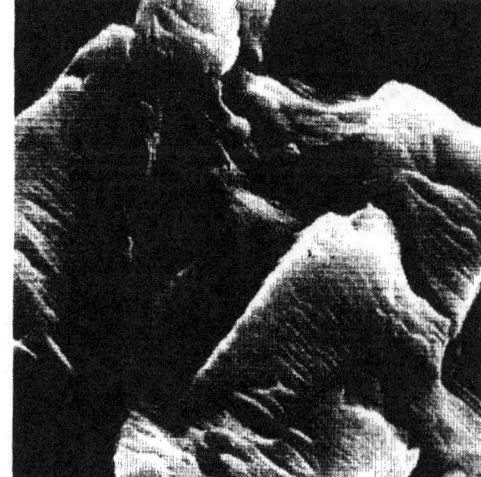

Fig. 5-41   10000X
Region 2.

Fig. 5-42   640X
Region 3. Fatigue striations.

ACKNOWLEDGMENTS

Many people contributed to this work. The author would like to thank the Materials Branch at Goddard Space Flight Center for making this work possible, particularly Jane Jellison, Michael Barthelmy, and Brad Parker for their technical and editorial advice; Diane Kolos for her metallurgical expertise, and Pat Friedberg for her diligent metallurgical specimen preparation. In addition, this author would like to thank Joe Generie for producing the CAD figures of the test specimens.

REFERENCES

1. American Society for Metals, *Metals Handbook*. 8th ed., Vol. 9, Fractography and Atlas of Fractographs, Metals Park, Ohio, 1974

2. S. Bhattacharyya, V.E. Johnson, S. Agarwal, and M.A.H. Howes, eds., *IITRI Fracture Handbook: Failure Analysis of Metallic Materials by Scanning Electron Microscopy*, IIT Research Institute, Metals Research Division, Chicago, 1979.

3. *1986 Annual Book of ASTM Standards,* Vol 3.01, Metals-Mechanical Testing; Elevated and Low-Temperature Tests, E647 and E399, ASTM, Easton, MD: 1986

4. *1986 Annual Book of ASTM Standards,* Vol 2.02, Die-Cast Metals; Aluminum and Magnesium Alloys, ASTM, Easton, Md: 1986

5. The Aluminum Association, *Aluminum Standards and Data 1984,* 8th ed., December 1984.

6. John E. Hatch, *Aluminum Properties and Physical Metallurgy,* American Society for Metals, Metals Park, Ohio: May, 1984

7. William F. Smith, *Structure and Properties of Engineering Alloys*, McGraw-Hill, New York: 1981.

8. American Society for Metals, *Metals Handbook*. 9th ed., Vol. 2, Properties and Selections: Nonferrous Alloys and Pure Metals, Metals Park, Ohio: 1979

APPENDIX 1

Material	Tensile Overload				High Cycle Fatigue
	293 K		77 K		
	Smooth	Notched	Smooth	Notched	
6061-T651 wrought	2	2	2	2	2
7075-T76 wrought	2	2	2	2	2
Ti-6Al-4 V wrought	2	2	2	2	2
CA172-TF00 wrought	2	2	2	2	2
6061- T651 welded	2	2	2	2	2

NUMBER OF SAMPLES TESTED IN EACH CONDITION

www.ingramcontent.com/pod-product-compliance
Lightning Source LLC
Chambersburg PA
CBHW081727170526
45167CB00009B/3727

* 9 7 8 1 5 0 2 9 8 9 4 6 8 *